Overcoming Productivity Challenges in Small Countries

Andre Haughton · Wendel Ivey

Overcoming Productivity Challenges in Small Countries

Lessons from Jamaica

palgrave
macmillan

Andre Haughton
Department of Economics
University of the West Indies
Kingston, Jamaica

Wendel Ivey
Bank of Jamaica
Kingston, Jamaica

ISBN 978-3-031-23300-5 ISBN 978-3-031-23301-2 (eBook)
https://doi.org/10.1007/978-3-031-23301-2

This Palgrave Macmillan imprint is published by the registered company Springer Nature Switzerland AG
The registered company address is: Gewerbestrasse 11, 6330 Cham, Switzerland

Dedicated to the generation of Jamaicans who will transform our hopes, visions and dreams to reality.

Foreword

This book on overcoming productivity challenges in Small Island Developing States (SIDs), with a focus on Jamaica, by Andre Haughton and Wendel Ivey, comes in the aftermath of the COVID-19 pandemic. However, for many decades now, the Jamaican economy has recorded anaemic rates of economic growth. There is an evident need to put the economy on a higher growth path, to reduce unemployment and poverty and raise the general standard of living of the Jamaican people. Productivity growth is a necessity for realizing these objectives.

Total factor productivity, which captures the overall efficiency of production, has declined at an annual average rate of 0.3 per cent since 1990. Similarly, labour productivity, measured by output per worker, contracted at an average rate of 0.3 per cent per year over the same period. Jamaica's labour productivity has lagged behind its major trading partners as well as a number of emerging market economies. Several factors have contributed to Jamaica's low productivity over the years. These include, but are not limited to, human capital deficiency, fiscal distortions and crime and corruption.

We now have a better understanding of these effects which are well-documented in this book. The authors also explain how important lessons from past events can help to forge a better way forward, not only for Jamaica, but also the rest of the Caribbean and Latin America.

The long-term impacts of events such as the pandemic on the Jamaican economy are still uncertain. This is especially true for productivity which only changes course in a significant manner over longer periods of time. This timely book from Andre and Wendel, highlights many key facets that could help policymakers respond to both short- and long-term challenges and make productivity a key part of Jamaica's long-term recovery.

Kemar Whyte
Principal Economist
The National Institute
of Economic and Social Research
London, UK

Preface

We all aspire to live in a country where there are equal opportunities for everyone to explore their full potential. An economy that allows the average citizen to earn a decent income, eat the best food, live in the dwelling of choice in peaceful communities that are governed by fair and practical institutions and rules of law. A country where people have access to good education, training, health care and social security and experience the best possible quality of life. Unfortunately, some countries are better able at providing this atmosphere than others for many current and historical reasons. Chief among them is the country's willingness, ability and efforts to enterprise itself to generate high income, which is rooted in their level of productivity.

Countries that are more productive, understand the importance of good supporting institutions and infrastructure to facilitate the continuous development of their economies in a positive spiral. In these productive developed countries, there is more to consume, more to save and/or more for people to invest. On the contrary, in lesser developed countries, people struggle with low productivity, low standard of living, poor health care, weak institutions coupled with depreciating infrastructure, sending the economy in a downward spiral. Here, there is less room to enable growth. Most of these citizens have to migrate to get the chance to earn a higher income or live a better quality of life. Even those with skilled labour and higher education often find it difficult to improve their standard of living in underdeveloped countries. These countries are often riddled by crime, violence and corruption. The vast majority are unskilled and tertiary-level education matriculation is low. Is it by design or social construct? Is it the mentality of the people, a lack of leadership, or is it sabotage? Or maybe a combination of all? Most developed countries are either from North America, Europe or Asia. While the Caribbean, Latin American and African countries are the least efficient, the least productive, and therefore find it hard to develop.

Historically, many of these countries have been victims of invasion, exploitation, slavery and genocide. Unfulfilled citizens are suppressed by political adversaries and economic ignorance. Jamaica is one such example. A former British colony that thrived on the backs of slaves that provided cheap labour to plantation owners who earn huge profits. After receiving independence in 1962, Jamaica has not overcome the trauma associated with slavery and its once productive slaves are now underutilized as free men. The Small Island Developing State (SID) located in the middle of the Caribbean basin with a population of about 2.9 million people, of which, roughly1.3 million are currently employed to various industries. Their efforts have to earn enough to take care of themselves and the other 1.6 million people who are not working formally. They complain about low wages and the cost of living. Jamaica, however, has a huge diaspora in other countries who send home high levels of remittances. The country generates most of its revenues from bauxite, tourism and remittance inflows, while spends most of its disposable income on mainly foreign goods. Outside of agriculture and light manufacturing, much of what Jamaicans consume (durables, nondurables and manufacturing final goods and inputs) are produced overseas. The increase in foreign demand promulgates a constantly depreciating currency which increases foreigners spending power in Jamaica but depletes Jamaicans spending power on foreign goods. A situation that should improve the terms of trade and has huge implications for the island's balance of payments. Imports are consistently three times the value of exports. Jamaica however has Net International Reserves (NIR) of over $3.5 billion USD enough to service its demand for foreign currency.

The United Nations have outlined the vision 2030 goals. A sustainability roadmap that most countries use as their benchmark target. Many countries have adopted these and must figure out how to achieve them along with their own. The objective is to make Jamaica the place to work, live and raise families by 2030. It appears rather impossible for Jamaica to achieve objectives in 10 years that it could not have achieved in 70 years.

The vision is closer to a dream than the reality. Jamaica, like other developing countries, struggles with innate handicaps that prevent them from achieving developmental objectives in a global environment characterized by climate change, increase polarization societies, increasing dependence on the internet and an ageing population. Jamaica's economy can serve as an example to other countries who are in similar positions.

This book explores Jamaica productivity challenges in its quest to become a developed nation by 2030. Chapter 1 puts things into perspective as to why it is important to understand and overcome productivity challenges in Jamaica and other small countries. It looks at the effect of and response to the COVID-19 pandemic and explores the new normal that has been propelled more rapidly by the advent of the virus. Chapter 2 explores the drivers of labour and total factor productivity using annual data in a Fully Modified Ordinary Least Squares (FMOLS) and Canonical Regression analysis. Here, corruption and human capital plays a significant role. Chapter 3 explores the use of labour in the goods producing industries and discuss their challenges, progression and future, while Chapter 4 explores the same for the service industries. Most of these industries appear to employ more people than what is required to fetch maximum productivity from the staff. Chapter 5 explores human capital, real wage and productivity and also the relationship between Compensation and Productivity in Jamaica. Productivity and compensation are found to be highly related, even more than in the United States. Chapter 6 conducts a survey to explore the implication of migration and human capital flight. Jamaicans migrate for better opportunities and to escape crime. Chapter 7 explores remittances and GDP growth that the country has been receiving from the people who migrate. Chapter 8 looks at corruption, causes and solutions. Chapter 9 does a comparative analysis between Jamaica and Mauritius; two similar countries but different approaches and different economic outcomes. Chapter 10 provides some alternative perspectives that are worth considering and charts a path forward.

Kingston, Jamaica

Andre Haughton
Wendel Ivey

Acknowledgement

We would like to thank our reviewers for taking time and extending the effort to review the manuscript. Their useful comments and suggestions are invaluable. We are particularly grateful to our research assistant Mr. Rameze Thorpe for his contribution to the book. We would like to thank our colleagues Ms. Sadé Morrison, Mr. Jonathan Isaacs, Mr. Simon Johnson and Mr. Julian Morrison for their helpful contribution to the discourse. We thank our families for their patience and support throughout this time of immense concentration and studying to properly execute this project. We also appreciate Princesses Aura, Aria and Ayla Haughton for their care and consideration. We would like to thank the Jamaica people for this useful experience to realize that productivity advancement strategies are idiosyncratic. The one-size-fits-all approach can be unfulfilling and therefore more energy has to be directed towards tailor fitting policies to suit individual country's needs.

Contents

About the Authors

Andre Haughton is Senior Lecturer in the Department of Economics at the University of the West Indies Mona with a research focus on International Economics and Macroeconomics. He is a former Senator in Jamaica where he reviewed acts and bills before passing into the law. Andre Haughton is an International Monetary Fund Distinguished Academic Fellow who earned his Ph.D. from the University of Essex in the UK. He is KPMG Peat Marwick, Thomas De La Rue and a British Commonwealth Scholar. He was selected as the University of the West Indies Most Outstanding Researcher in 2017 for the Faculty of Social Sciences and the Most Outstanding UWI Alumni for the decade 1999–2008. He is the winner of Allied Academics best research paper award in 2012.

Wendel Ivey is currently an Economist at the Bank of Jamaica in the Research and Economic Programming Division where he conducts Labour Market and Potential Output Analysis and forecasts. Prior to working at the Bank of Jamaica, he was a Senior Research Analyst in the Research and Measurement Unit at the Jamaica Productivity Centre where he conducted productivity analysis, calculations and benchmarking. Wendel's research interest is on economic development particularly with a focus on productivity. He holds a Master of Science in Economics with a specialization in Economic Development and International Economics. He also holds a Bachelor of Science in Economics and Statistics. While doing his Masters, he finished in the top 2 of the

Most Outstanding Student Researchers at the 97th Annual Meeting of the Southwestern Social Science Association (SSSA) held in Austin, Texas. Wendel also finished in the top 3 of the International Monetary Fund (IMF) Caribbean Essay Competition on Inclusive and Resilient Growth in the Caribbean region.

Acronyms

ACB	Anti-Corrution Branch
AFF	Agriculture, Forestry and Fishing
AIC	Akaike Information Criterion
ARDL	Autoregressive Distributed Lag
ASEAN-5	Association of Southeast Asian Nations Five
BOP	Balance of Payments
BPO	Business Process Outsourcing
BRICS	Brazil, Russia, India, China and South Africa
CCR	Canonical Cointegrating Regression
COMESA	Common Market for Eastern and Southern Africa
COVID-19	Coronavirus Disease
CPT	Care and Protection Tax
CUSUM	Cumulative Sum or Recursive Residuals
CUSUMSQ	Cumulative Sum of Squares of Recursive Residuals
DPP	Directorate of Public Prosecution
ECB	European Central Bank
ECI	Economic Complexity Index
ECOWAS	Economic Community of West African States
ECT	Error Correction Term
EFF	Extended Fund Facility
EPZ	Export Processing Zone
EU	European Union
FDI	Foreign Direct Investment
FMOLS	Fully Modified OLS
FMOLS	Fully Modified Ordinary Least Squares
GATT	General Agreement on Tariffs and Trade

GCI	Global Corruption Index
GDP	Gross Domestic Product
GETS	General to Specific
GNI	Gross National Income
GSAT	Grade Six Achievement Test
GSS	Global Services Sector
HC	Human Capital
HDI	Human Development Index
ICAC	Independence Commission against Corruption
ICLS	International Conference of Labour Statisticians
IMF	International Monetary Fund
ITA	Invitation to Apply
JAMDEX	Jamaica Digital Exchange
JLP	Jamaica Labour Party
JMMB	Jamaica Money Market Brokers
JPC	Jamaica Productivity Centre
JRC	Jamaica Railway Corporation
JSE	Jamaica Stock Exchange
JUTC	Jamaica Urban Transit Company
LAC	Latin America and Caribbean
LNG	Liquefied Natural Gas
LP	Labour Productivity
MDA	Ministries, Departments and Agencies
MENA	Middle East and North African
MLSS	Ministry of Labour and Social Security
MOCA	Major Organised Crime and Anti-Corruption
MSME	Micro, Small and Medium Enterprises
NAFTA	North American Free Trade Agreement
NIA	National Integrity Action
NPO	National Productivity Organization
OEC	Observatory of Economic Complexity
OECD	Organization of Economic Cooperation and Development
P.A	Per Annum
PCA	Principal Component Analysis
PCJ	Petroleum Corporation of Jamaica
PEP	Primary Exit Profile
PNP	People's National Party
PPF	Production Possibility Frontier
PSOJ	Private Sector Organization of Jamaica
R&D	Research and Development
SBC	Schwarz Bayesian Criterion
SCI	Sygnust Credit Investments
SDGs	Sustainable Development Goals

SIDS	Small Island Developing States
STATIN	Statistical Institute of Jamaica
TEF	Tourism Enhancement Fund
TFP	Total Factor Productivity
UK	United Kingdom
UN	United Nations
UNDP	United Nations Development Programme
US	United States
WB	World Bank
WHO	World Health Organization
WRTRM	Wholesale and Retail Trade, Repairs: Installation of Machinery and Equipment
WWII	World War II

List of Figures

List of Tables

CHAPTER 1

COVID-19 and Efficiency in a New Global Economy

1.1 Why Is Productivity Growth Important?

Jamaica and many other countries in the Latin America and the Caribbean (LAC) have grown and developed at a slower pace relative to other countries in other regions around the world due to low levels of labour and total factor productivity, rather than due to a lack of investment (Daude & Fernández-Arias, 2010).[1] Among all the regions, East Asia has experienced the largest productivity growth of 816 or 11.7% per annum (p.a) between 1950 and 2019. Central and Eastern Europe has witnessed the second highest productivity growth of 458% (6.5% p.a.), Western Europe 357% (5.1 p.a), North America 219% (3.1% p.a), Oceania 165% (2.4% p.a), while Latin America and the Caribbean has experienced the second lowest productivity growth of just 146% (2.1% p.a) outside of Africa whose productivity grew by only 122% (1.7% p.a) over the past 70 years, see Table 1.1. Higher productivity growth has enabled the countries from North America, Western Europe and Eastern Asia to accelerate their Gross Domestic Products (GDPs) and transition their economies from developing to developed, see Ball et al. (2016), Beugelsdijk et al. (2018), Chen (1997), Felipe (1999), Lee and Xaun (2019), Myrdal (1972), Ngo & Nguyen (2020), Timmer et al. (2011), and Van Der Eng (2009)

[1] A common misconception is that many SIDS fail to develop due to a lack of investments.

A. Haughton and W. Ivey, *Overcoming Productivity Challenges in Small Countries*, https://doi.org/10.1007/978-3-031-23301-2_1

Table 1.1 Global productivity growth by region

Region	*1950–1990 (%)*	*1990–2019 (%)*	*1950–2019 (%)*
Africa	70.0	31	122
Asia	311.1	123	816
Central and Eastern Europe and Central Asia	193.5	90	458
Latin America & the Caribbean	85.7	32	146
The Middle East	−48.5	−18	−58
North America	119.9	45	219
Oceania	89.7	39	165
Western Europe	228.1	39	357
Average	131.2	47.7	278.0

Source Conference Board

Jamaica, like many other Small Island Developing States (SIDS), wants to accomplish its United Nation's Vision 2030 Sustainable Development Goals (SDGs) to become the country of choice to live and raise families. This is possible if the country can increase its Gross Domestic Product (GDP) through increased productivity as reasoned by Daude and Fernández-Arias (2010). Increasing GDP and productivity has been a challenge for the island in a global environment where productivity growth has been slowing down, see Fig. 1.1. Van Ark (2016) posits that productivity growth has been declining globally since the mid-2000s. The UN (2017) report on productivity also suggested similar evidence. The United Nations outlined that many SIDS might not be able to achieve their SDGs by 2030 due to sluggish economic performances. Like other SIDS, Jamaica is import-dependent, small, price-taking, and susceptibility to external shocks such as global geopolitical occurrences, natural disasters and climate change. These have impacted the GDP output and developmental process of the economy negatively over the years. Domestic hindrances such as high levels of crime, corruption and institutional weaknesses have also prevented the nation from building its resilience and overcoming developmental challenges. Therefore, for Jamaica and other SIDS to transition to high-income economies, a conscious decision must be made to prioritize factors that can enhance productivity, which is the foundation of economic growth and progression. Productivity growth is also an important factor in business cycle fluctuations, monetary policy prescriptions, inflation, exchange rate and

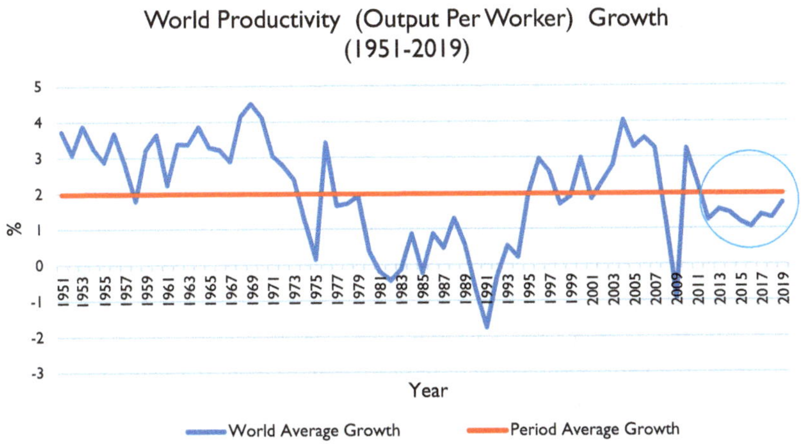

Fig. 1.1 Global productivity growth (1951–2019) (*Source* Conference Board, authors' compilation)

other key macroeconomic variables such as investments, consumption and employment (Gomez-Salvador et al., 2006).

Productivity in its most basic form can be expressed as how much output can be produced per unit of input employed within any process. Two types of productivity measures dominate the economic literature: Total Factor Productivity which measures how well capital, labour and the working environment combine to produce output (the Solow 1956 Residual).[2] It accounts for factors such as technology, managerial quality and the environment among others. Labour productivity is defined as output per worker employed or output per hours worked. In 2020, Labour productivity in Jamaica was roughly $20,000, the same as it was back in 1958, see Fig. 1.2. Growing at an average rate of 2% per annum, while across the Caribbean, over the same period, labour productivity grew more rapidly; in Barbados, productivity grew by 89% to $40,000, Trinidad and Tobago by 148% to $69,000, the Dominican Republic by 188% to $41,000 and St Lucia by 217% to $33,000, see Table 1.2 and Fig. 1.4. Jamaica experienced its most productive era and highest

[2] Cobb Douglas Function.

economic growth spurt more than 50 years ago when labour productivity trended upwards to more than $34,000 US per person in 1973, see Fig. 1.2. GDP per capita grew by as much as 11% in 1970 and peaked at $5,600 USD in 1972, see Fig. 1.3 after which they trended downwards, and the island has not materialized that level of growth and productivity since then.

Comparatively, over the same 70 years, labour productivity in the Asian-5 countries have grown at a more rapid pace than any other region. The Asian-5 countries have grown by more than 1000% compared to 271% growth on average in the Caribbean top 5 nations, 5 times more, see Table 1.2. Among the emerging Asian countries, Taiwan and Thailand have seen the most productivity growth but Singapore and Taiwan have the highest labour output valued at $155,000 and $111,000 per annum respectively, see Table 1.2. In 2019, Taiwan's labour productivity was valued five times that of Jamaica's and Singapore's that was on par with Jamaica in the 1950s is now valued 7 times the labour productivity of Jamaica, see Table 1.2. Indonesia has one of the lowest labour outputs per person among the Asian top 5, valued at $29,000, growing by more than 480% since 1950, and is 1.5 times the labour productivity of Jamaica, see Table 1.2. Such a huge disparity in productivity growth in Jamaica relative

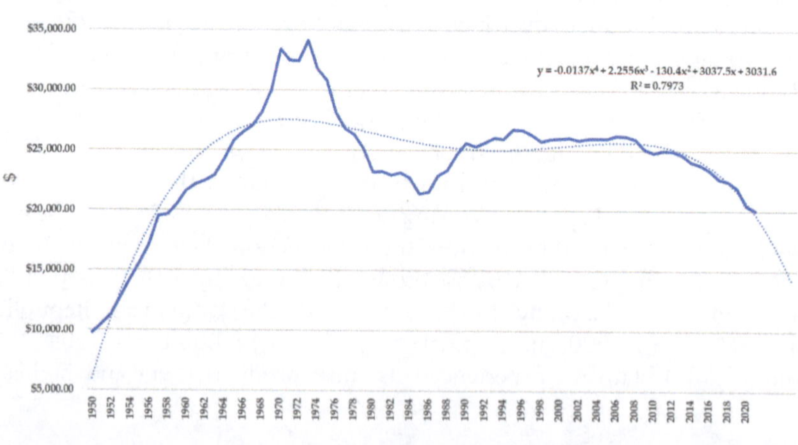

Fig. 1.2 Jamaica's output per employed person (1950–2021) (*Source* Conference Board)

Table 1.2 Labour productivity growth/output per worker in Asia and the Caribbean (1950–2019)

ASIA	*1950*	*Growth 1950–1990 (%)*	*1990*	*Growth 1990–2019 (%)*	*2019*	*Total growth 1950–2019 (%)*
Indonesia	$4,931	141	$11,860	145	$29,035	489
Malaysia	$10,535	204	$32,077	116	$69,288	558
Singapore	$19,511	317	$81,337	90	$154,946	694
Taiwan	$4,237	917	$43,078	159	$111,466	2531
Thailand	$2,964	357	$13,548	166	$35,979	1114
Average growth		387		135		1077
The Caribbean						
Barbados	$21,262.72	95	$41,370.48	−3	$40,170.35	89
Dominican Republic	$14,287.87	37	$19,570.53	110	$41,188.71	188
Jamaica	$9,640.20	145	$23,654.56	−15	$20,076.14*	108
St. Lucia	$10,466.80	208	$32,197.81	3	$33,163.15	217
Trinidad & Tobago	$27,852.03	18	$32,925.56	110	$69,205.25	148
Average growth		76		41		129

Source Conference Board; *illustrate the country with the lowest level of productivity growth

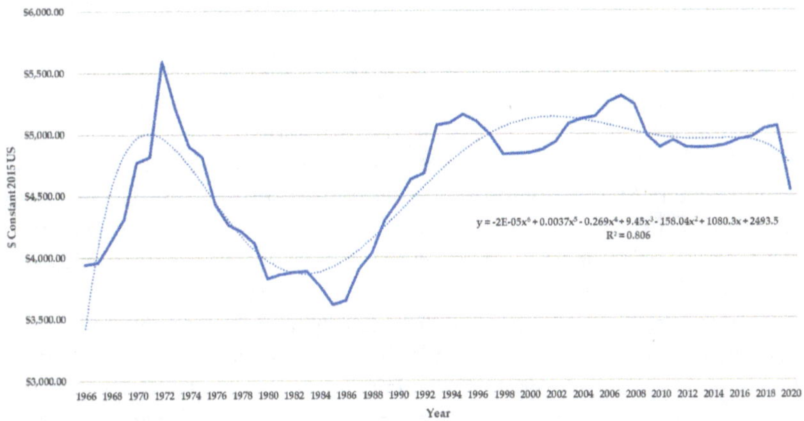

Fig. 1.3 Jamaica's GDP per capita (1966–2020) (*Source* World Bank)

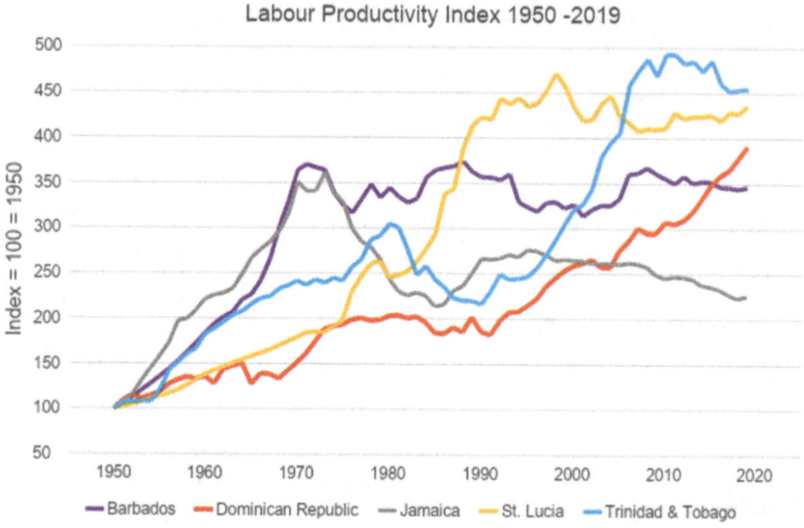

Fig. 1.4 Labour productivity in the Caribbean (*Source* Conference Board)

to the other countries in the Caribbean and Asia is a cause for concern and must be addressed, making Jamaica a useful case study to analyse the causes of and solutions to the nation's poor productivity growth. Why is it that these other countries with similar economic history and innate characteristics as Jamaica have experienced higher productivity growth and economic development than Jamaica has?

Timmer and Los (2005) along with Hsieh (1999) suggests that much of the rapid growth and industrialization of East Asian countries in the second half of the twentieth century occurred through advancements in technology and innovations. Gomez-Salvador et al. (2006) also found similar evidence to suggest that government policy directed towards improving factors like technology can enhance productivity in the Euro area. Crespi and Pianta (2008), Geroski (1989) and Uppenberg and Strauss (2010) concur. Many SIDS including Jamaica have found it challenging to properly incorporate technology and innovation into their productive processes and as a result, have not yielded increased productivity and high output growth. In Jamaica for example, the economic environment has not been enabling to technologically driven infrastructure and as a result, it has been slow to be absorbed. For instance, at least

15% tariff plus other import duty are levied on computers and similar technological devices and equipment. These tariffs artificially inflate the cost of these items and reduce the technology absorption rate of the country. Much of Jamaica's industrial policy subsidies have been directed towards industries like tourism and agriculture in a non-innovative or non-technologically advancing manner. Both industries are highly volatile, low growth, low value-added and low productivity industries. There is no real innovation occurring in these industries and no other fast-growing industries to facilitate the rapid growth of GDP. This has been so for many other SIDS as well. If Jamaica and other SIDs want to achieve rapid growth like the Asian-5, they must innovate and make it cheaper to incorporate technology in industries with the potential for high growth.

Using industrial policy to drive innovation and incorporate technology in the economic productive process has been challenging for many SIDs including Jamaica, who either lack the know-how or the will to do so. Successive governments pitch their policy prescriptions in their annual budget presentations.[3] These proposals have had varying effects on different industries but most, if not all, have been unfulfilling. The usual excuse appears to be that politics rather than economics predominantly determine the industries targeted and the policy orientation in these small countries. Politics however exist in every country, but there appears to be a culture in Europe, North America and Asia that has allowed their political decisions to coincide to a large extent with their economic aspirations which have borne fruit. One plausible argument is that more checks and balances are present to increase the accuracy of government decisions and to reduce the squandering of public resources. Despite incidences of corruption, these countries have used Industrial Policy productively to innovate and industrialize.

One other compelling argument is also the role of politics in policy continuation. Different political parties with different ideologies have different approaches to economic management and policy orientation. Many times, developmental objectives initiated by one party are sidelined by the other party when they demit office. In these instances, a change of government can change the policy orientation of the country entirely. In more advanced countries such as the United States and the United Kingdom or Canada, a blueprint is laid out in stone to guide

[3] These can be found in the budget presentations of the Minister of Finance over the years and can be found on the Ministry of Finance website at https://mof.gov.jm/.

their economic trajectory. Irrespective of the political party in power, the national objectives and a culture of growth are maintained and pursued. In lesser developed countries like Jamaica and other SIDs, there are declared national agendas like the UN Vision 2030 SDGs but the political parties' approaches are incongruent. This disrupts the crucial policy continuation that is needed for successful industrial policy. This lack of policy continuation was recognized by the International Monetary Fund (IMF), who ensured that measures were put in place to rectify this before the latest Extended Fund Facility (EFF) was granted to Jamaica.

A UN Report 70 years ago highlighted that SIDS face three confounding challenges in their developmental pursuits; (1) the ability to increase the productivity of their labour force, (2) the ability to increase the capacity to generate savings and (3) The ability to increase external real purchasing power. These challenges are still relevant today. Myrdal (1972) outlines that there are 6 aspects of the economy that are interrelated and must be understood and addressed simultaneously to enable a country to achieve holistic economic development. (1) output and income, (2) conditions of production, (3) standard of living, (4) attitude towards life and work, (5) Institutions and (6) Policies. The first two according to Myrdal (1972) are normally the focus of economies because they provide income and wealth but must be supported by the last four if the country is to progress. Collectively, these factors contribute to a country's level of productivity. Productivity was seen as the most perplexing factor that hindered the economic development of Asian countries in the middle of the twentieth century (Myrdal, 1972). Today, many of these nations have high growth, high income, developed and fairly stable economies.

The aim of this book is to put Jamaica's developmental pursuits under the microscope and to address the issue of productivity, which is essential to increasing economic growth and achieving long-term economic development. This chapter will shape the discussion. The rest of the chapter is outlined as follows: Sect. 1.2 discusses the Jamaican economy prior to the COVID-19 pandemic and its impact on the country. Section 1.3 looks at the world's thrust into the Fourth Industrial Revolution (4IR) as a result of the new normal caused by the lockdown measures.

1.2 COVID-19 and the Jamaican Economy

Prior to the COVID-19 pandemic in 2020, Jamaica was on the verge of making a significant turnaround from the high debt and low growth conundrum they experienced in 1990s and 2000s. Economic reform aided by an Extended Fund Facility (EFF) by the International Monetary Fund (IMF) in 2013 as well as assistance from Inter-American Development Bank (IDB), the World Bank (WB) and other institutions enhanced the island's pursuits. This translated into improvements in some of the island's major macroeconomic variables which started to trend in the direction of a nation on the verge of economic take-off. Interest rates trended downwards from more than 30% in the 1990s to less than 1% in 2019. Inflation trended downwards from more than 20% per annum in the 1990s to less than 6% in 2019. Net International Reserves (NIR) grew to more than US $3 billion in 2019 from negative in 1990. Infrastructural development was expanding especially in real estate and road construction. By 2015, the Jamaica Stock Exchange (JSE) was among the best performing stock markets in the world. It continued to trend upwards, see Fig. 1.5. Moody's and S&P upgraded the island's credit ratings for the first time in decades. Jamaica was also able to lift approximately 124,000 people (4,575 people per year) out of poverty over the last 3 decades (1990–2017).[4]

The local debt, one of the focuses of the economic reform fell from 146% of GDP in 2013 to 96% by 2019. The country was on a steady path to reduce its debt to GDP ratio to 60% by 2025. Lower debt would create room in the budget for more spending on capital, infrastructural expansion, education, health care and social services. GDP growth however was not accelerating above its usual 0.7% per annum and Jamaica has had the second lowest GDP growth on average in the Caribbean between 2000 and 2019, see Fig. 1.8, and the 4th lowest GDP per capita among Caribbean countries in 2019, see Fig. 1.9. The island's Gross Domestic Product (GDP) grew at a slower pace after the Global Financial Crisis than it did throughout the 90s and early 2000s. As usual, Jamaica has had its original challenges, a constantly depreciating currency and high dependence on imports especially energy. Riddled with corruption and high crime and violence, Jamaica is consistently ranked within the top 5 in the world in terms of deaths per hundred thousand where gang violence and

[4] Authors' calculation using population and poverty data.

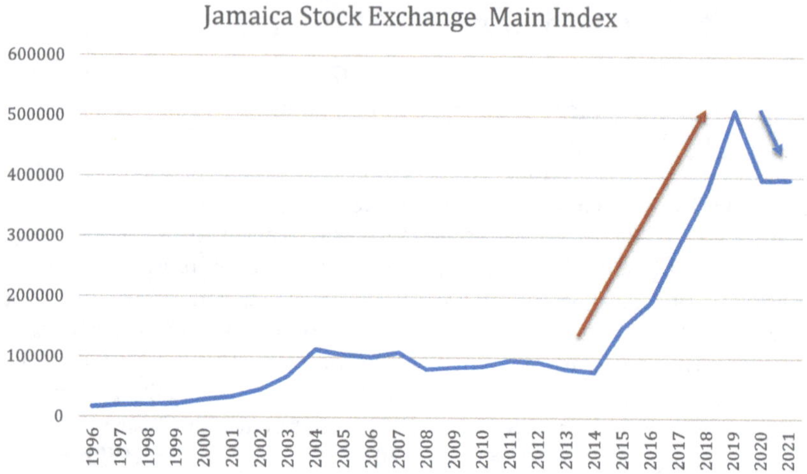

Fig. 1.5 JSE main market index (*Source* Jamaica Stock Exchange)

robberies continue to deter business activity. Low wages and ambiguous labour contracts without proper benefits limit the standard of living of the labour force as large proportion of employed workers earn less than US $10,000 per year (approximately US $800 per month) excluding people who are self-employed. The workforce is also not properly equipped with the requitable skills nor knowledge to consistently produce high value-added goods and services that will help Jamaica fetch higher revenues in the global marketplace and improve their quality of life. Jamaica's score of 7.35 on the Human Development Index (HDI) in 2020 is the same as it was in 2010, no improvement.

In 2020, Jamaica qualified as an Upper Middle-Income country with a Gross National Income (GNI) per capita of $4,670 US, declining from $5,250 in the years prior, see Fig. 1.6. This was partly due to the COVID-19 pandemic which influenced a 11.0% decline of Jamaica's GNI, more than twice the decline witnessed during the Global Financial Crisis in 2008, see Fig. 1.7. The island is in danger of regressing even further to the Lower Middle-Income category if its GNI per capita continues on the same downward trend and falls below the lower threshold of $4,053, see Fig. 1.6. The country is therefore challenged with the task of reorienting the trajectory of its economy. Firstly, to regain its 2019 levels of GNI per capita. Secondly, to triple it, if it is to jump above the upper threshold of $12,695 US required to be classified as a High Income Developed Nation and thirdly, to achieve its vision 2030 SDGs. As a result of these challenges, there has been a mass migration of the most skilled and highly educated Jamaicans to more developed nations. These and other challenges faced by Jamaica have been exacerbated due to the onset of COVID-19 pandemic.

The COVID-19 virus was profound; it resembled the Spanish Flu of 1918/1919. A deadly virus spreading rapidly with no immediate cure. Governments around the world ordered curfews and lockdown measures

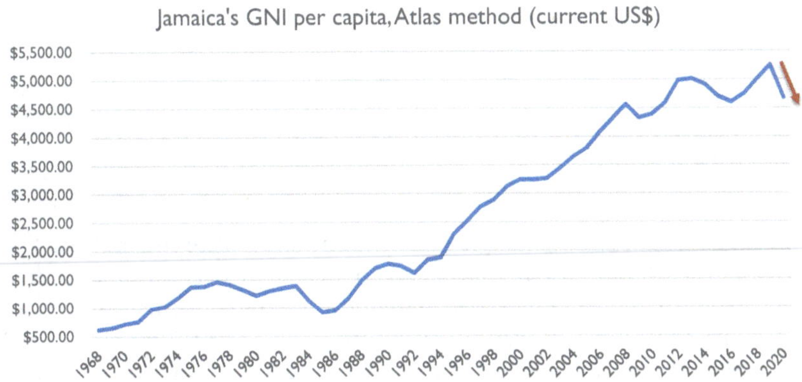

Fig. 1.6 Jamaica's GNI per capita (1968–2020) (*Source* World Bank)

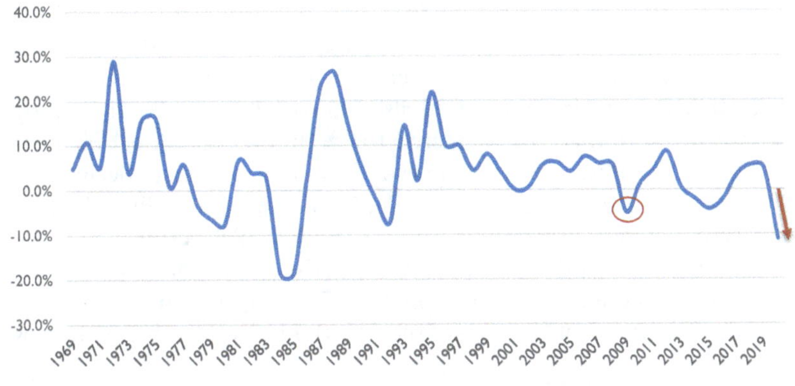

Fig. 1.7 Jamaica GNI per capita growth (1969–2020) (*Source* World Bank)

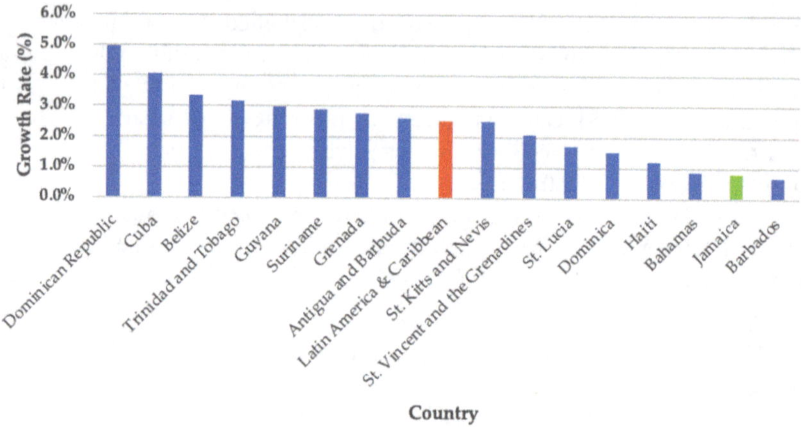

Fig. 1.8 Average annual growth of Caribbean economies (2000–2019) (*Source* World Bank)

as first responses, forcing people to stay home and isolate themselves. The initial effect on movement, travel and tourism spilt over to other industries. More than 200 large companies and thousands of small businesses had either shut down or filed for bankruptcy in the first 6 months of the pandemic. It had an immediate negative impact on production, global logistics, commerce, finance and distribution channels.

The negative economic impact of COVID-19 was worse than the Global Financial Crisis (GFC) of 2008/2009. The Jamaican economy declined by 10% or $200 billion JMD in 2020, more than 4 times as much as it was in the 2008/2009 GFC, see Fig. 1.10. In 2019, Jamaica had one of the lowest per capita incomes in the Caribbean region, see Fig. 1.10. GDP per capita in 2020 also fell to 1990s level and Employment fell by more than 4% compared to 1% during the Global Financial Crisis.[5] The Caribbean Gross Domestic Product on average declined by 4.2%, Central America declined by 7.1%, South America declined by 6.6% and North America also declined by 4% in 2020. All performing worse than they did back in the Global Financial Crisis. In Jamaica, for the year 2020, labour productivity fell by more than 5% compared to falling by just 1% in the Global Financial Crisis, see Fig. 1.11. By March 2020, most

[5] See Fig. 1.3.

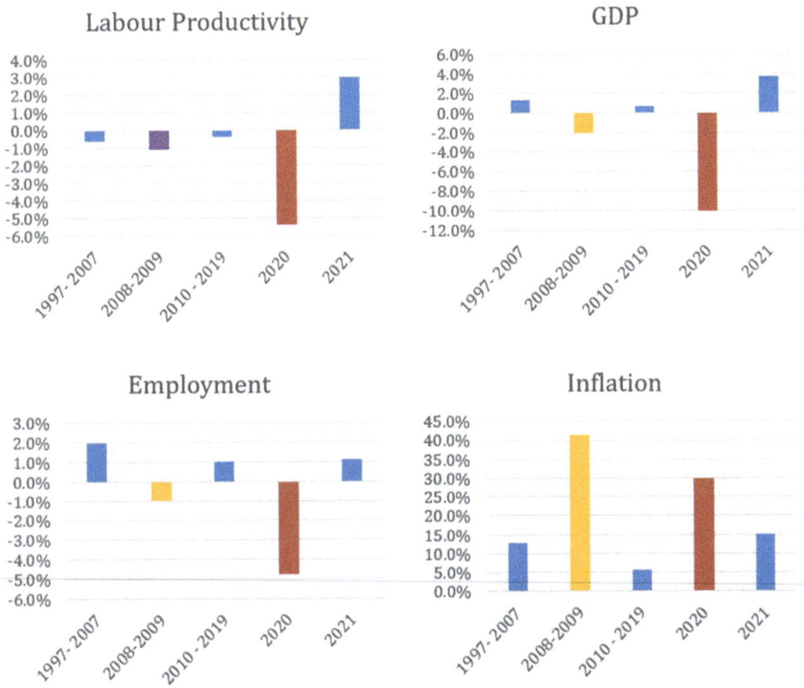

Fig. 1.9 Pre, during and post GFC and COVID-19 crises change in labour productivity (top left). GDP (top right), employment (lower left) and inflation (lower right) (*Source* World Bank)

of the major stock exchanges around the world entered bearish trends. Locally, the Jamaica Stock Exchange (JSE), which was named the best performing in the world in 2017, declined by more than 17% due to COVID-19, compared to 6% in the 2008/2009 GFC, see Fig. 1.12. The government responded by spending beyond what was fiscally prudent. Delaying Jamaica's debt objective by reversing the trajectory of its debt to GDP ratio; increasing from 96% of GDP in 2019 to 110% in 2021. Many other countries witnessed similar circumstances and needed assistance to rejuvenate. More than 80 countries sought emergency financial support from the IMF and more loomed as the virus spread and employees stayed away from physical work to contain the pandemic. Global production and productivity needed ransom.

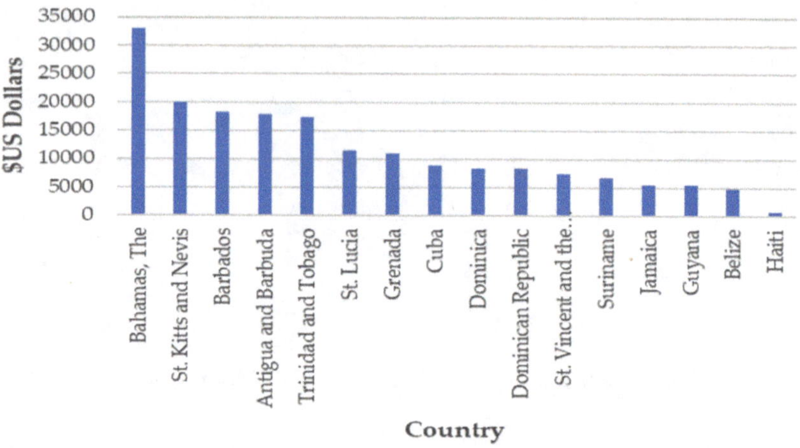

Fig. 1.10 GDP per capita in the Caribbean (2019) (*Source* World Bank)

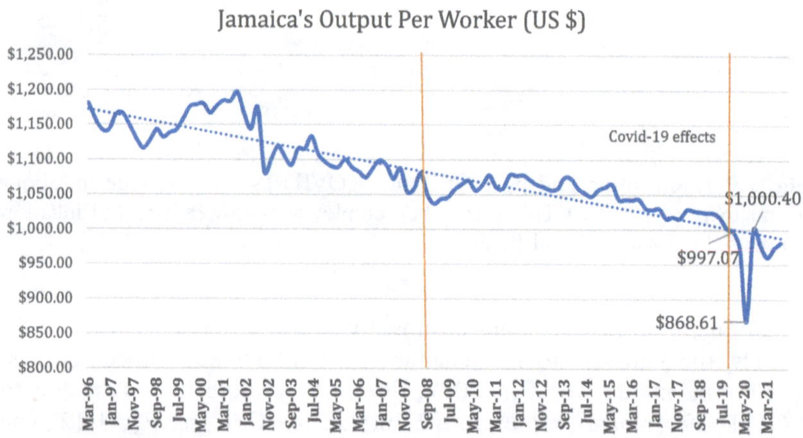

Fig. 1.11 Output per worker (US $) (*Source* STATIN)

The global economic downturn instigated by the COVID-19 shock will further delay many developing countries' Vision 2030 Sustainable Development Goals (SDGs). Jamaica and other SIDs are therefore tasked with the job of coping with the challenges arising from the pandemic, as well as, to figure out how to best overcome the longstanding innate

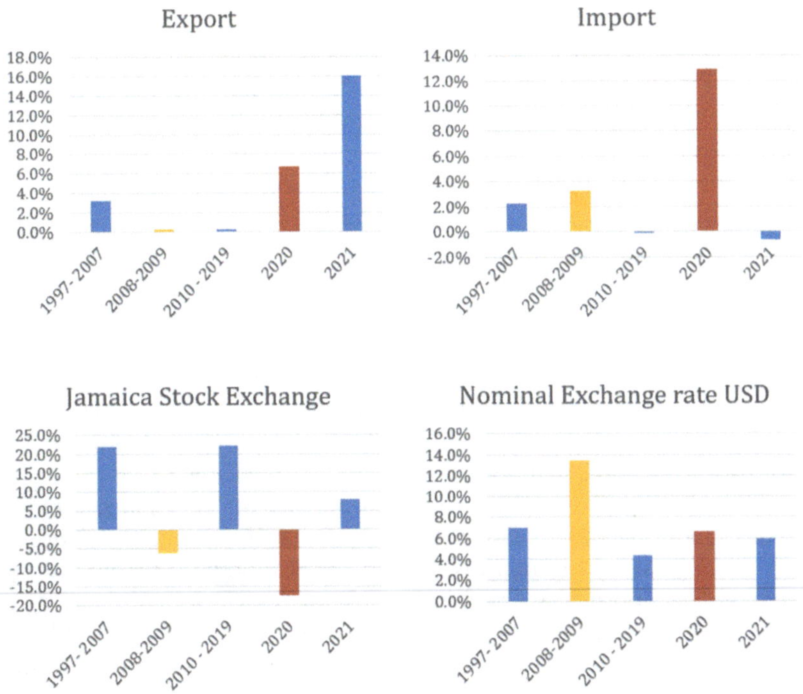

Fig. 1.12 Pre, during and post GFC and COVID-19 crises change in export (top left). import (top right), Jamaica stock exchange main index (lower left) and nominal exchange rate (lower right) (*Source* World Bank)

structural and other handicaps that have restrained the productivity of their labour force and suppress the output of their economies, which has kept a significant portion of their population in poverty (Guillaumont, 2013).

Correcting economic depressions usually depend on the government or central authorities' policies directed at minimizing the impact of the crisis and restoring stability. Normally in a demand suppressed atmosphere, for example, recall the harsh economic conditions of the great depression of the 1930s, Classic expansionary fiscal policy recommended by Keynes (1937) proved useful. These polices were directed towards expanding infrastructure and facilitated the expansion of industries and logistics which increased economic activity. At the same time, workers earned income which restored aggregate demand to normal levels, allowing the United States and the world to restore stability and rebound. But the COVID-19 crisis was different. It was like the Great Depression and the Spanish Flu combined. A health and economic crisis infused together. What made the situation worse was those measures to reduce the health effects, increased the economic strain and vice versa. Any fulfilling recommendation to escape the crisis had to balance both.

Fundamentally, the avenue through which the fiscal and monetary policy tools enter the real economy is crucial to facilitate the effectiveness of any stabilization policies or strategies administered. Learning from Keynes (1937), contemporarily, many governments attempted to counterbalance the COVID-19 lockdown with fiscal stimulus packages directed to the neediest and the most vulnerable in their countries. Initially, Japan redistributed about 21% of its GDP back to its citizens, the United States gave back 13%, Sweden 12%, Germany 10.7%, France 9.3%, Spain, Italy, the UK, China and South Korea round out the top ten. Jamaica redistributed about 3% of its GDP. The United States, for example, exhausted their entire budget eight months into the first year of the crisis and needed congressional approval to spend more. These excess spending were to boost aggregate demand while people were out of work during the lockdown. However, for many countries, reality did not play out like Keynes and the Great Depression because there was a fundamental difference. Most of the COVID-19 stimulus packages were consumption-driven and brought about no noticeable changes in the capacity of a country to produce. Mostly just one-way transfers that are not calculated in GDP.

More income from the stimulus packages increased global demand which depleted inventories. Additionally, the pandemic displaced workers, reduced production and clogged logistical channels. Apple, for example, reported that it lost $6 billion in sales due to shortages. There were also shortages and delays for many other goods for companies which increased the cost of shipping and delivery. This coupled with rising oil prices led to higher inflation. Global inflation began to increase in 2021, threatening the inflation target of many countries, putting many governments and central banks on edge as to how to respond. The US 12-month inflation rose to 6.2% in October 2021, the highest in 30 years. Inflation in the UK, EU, BRICS, Latin America, Asia, Oceania, Africa and the Caribbean rose above predicted levels. In Jamaica, inflation rose to a higher level relative to the Global Financial Crisis, see Figs. 1.9 and 1.13. Higher inflation depletes real wages and reduces the standard of living. Countries had to implement polices to counteract the rising prices. Brazil and the UK were some of the first to respond by increasing interest rates to curtail aggregate demand. Jamaica increased interest rates first from 0.5 to 1% then to 4% in one month. This in theory, based on the Taylor Rule can curtail inflation but given that it's a contractionary policy, it also drags down GDP as well. Some central bankers were slow to react, citing that the inflation was transitory and need not intervene. These economies needed to revive and quickly but as the virus prolonged the policy response became trickier.

Mass vaccination was necessary to restart these economies. There was a massive vaccination campaign globally and locally. In Jamaica, the first 50,000 doses of vaccines were received on the island on March 8, 2021, donated by the Government of India. The vaccination process began soon after on March 10, 2021, exactly one year after the first COVID-19 case was confirmed in Jamaica. Another 14,400 vaccines were received through the Covid-19 Vaccines Global Access (COVAX) facility one week later March 15, 2021. Many more vaccines were imported to the island, much of which had to be dumped due to breaches of expiration dates because of vaccine hesitancy by the working class. On September 30, 2021, roughly 55,000 doses of vaccines were dumped due to non-take-up. Notwithstanding, the local vaccine drive continued as celebrities encouraged people to take the vaccine. Local and international debates ensued as to whether governments should implement mandatory vaccine orders. In Jamaica, the stance was supported by the Private Sector

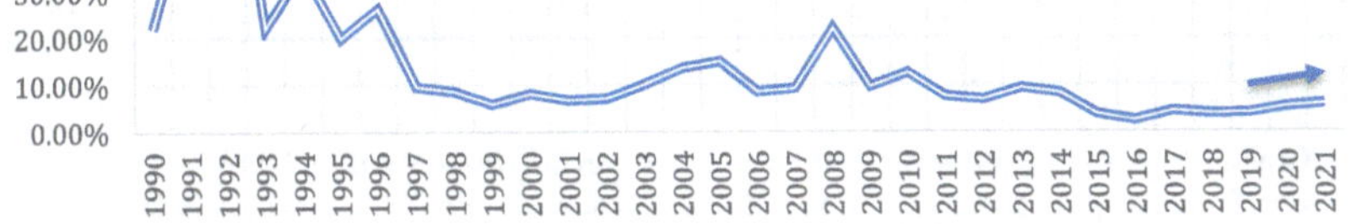

Fig. 1.13 Jamaica inflation rate (*Source* Statistical Institute of Jamaica)

Organization of Jamaica (PSOJ). Two of the island's largest employers, the largest bank and second largest telecoms provider issued mandatory vaccine orders to their staff. These were later challenged and overruled by the respective workers' unions as many of their staff were hesitant to take the vaccine. Official vaccination data shows that up to December 30, 2021, only 19% of the Jamaican population was fully vaccinated and 23.4% up to May 24, 2022.[6] A low vaccine take-up meant a slow return to the workplace. But this might not be all bad.

1.3 Infusion of Innovation and Technology in Production Processes and Linkages

The world's thrust into the Fourth Industrial Revolution has been further accelerated by the COVID-19 pandemic. Work from home programmes that were originally intended as short-term measures to minimize the spread of the virus, turned out to be long-term solutions by enabling some businesses to become more efficient. Many firms realized lower overhead costs due to more of their staff working remotely. Many of whom will be unwilling to revert to the more expensive face-to-face ways of doing business after the pandemic ends. Some employees have expressed their willingness to work from home at least one day a week after the pandemic has ended. A survey conducted by The Economist asked managers if remote work made their employees more, less or equally as productive. More than 40% of the managers from the United States responded that remote work made employees more productive compared to roughly 30% who said remote work made their staff less productive. The numbers were similar in Switzerland, Britain and Chile. But not for Russia, Germany, Canada and Belgium where more managers reported that remote work made employees less productive. Nevertheless, there are some companies that have been and will continue to benefit from remote work.

More people being at home has reshaped the world, a phenomenon referred to as the new normal. The long-term economic paradigm

[6] Covid-19 Vaccine Official Statistics taken from OURWOLDINDATA. https://ourworldindata.org/explorers/coronavirus-data-explorer?zoomToSelection=true&time=2020-03-01..latest&facet=none&pickerSort=asc&pickerMetric=location&Metric=People+fully+vaccinated&Interval=Cumulative&Relative+to+Population=true&Color+by+test+positivity=false&country=~JAM.

shift from mainly face to face to virtual or hybrid interactions has expanded new avenues for new jobs, like bloggers, social media managers, marketers, etc. Many companies and countries are preparing for this new normal and have been putting structures in place to fully capitalize on the transition. Social and economic empowerment platforms allow better access to markets for goods, services and talent. Amazon, the largest eCommerce company, has extended their services to Jamaica and other SIDs to sell goods and services on their website. They reported a net income of US $33.36 billion in 2021 twice as much as Jamaica's GDP. Web 3 has allowed value to be transferred via the internet in a decentralized manner. Cryptocurrency, blockchain technology, payments systems, applications and automation are expanding and will affect how we live, think and survive. Bitcoin, Ethereum and other alternative coins' high level of price volatility pose a threat to mainstream acceptance and usage.

Digital currency, block chain technology, payments systems, applications, automation and transportation usage had begun to expand. El Salvador was the first country to establish bitcoin as a legal tender although they face challenges as it relates to the implementation and lack of assistance from global financiers such as the World Bank.[7] Their pioneering experiment is a pilot study for a world that is familiar with a less volatile medium of exchange. Low-valued currency, lack of global financial intermediaries, low financial inclusion and financial challenges in small countries create an atmosphere to experiment with plausible solutions. This poses a problem to some central bankers who are accustomed to being in control of the distribution of money in their jurisdictions. Nevertheless, cryptocurrencies are a global phenomenon and anyone with access to the internet can have access to them. To many, this seems like a reasonable solution to minimizing the transaction cost and waiting time associated with local and international money transfers in SIDS. As well as, to increase financial inclusion for the unbanked in remote rural areas in poor countries.

Despite the common hesitancy, more governments are buying into the idea and are experimenting with their own alternative coins and bonds, etc. In 2021, the European Central Bank (ECB) issued a bond on the Ethereum platform. This might open the mindset for mainstream acceptance of blockchain technologies and cryptocurrencies. Jamaica and many

[7] The World Bank Can't help. But the IMF Might help El Salvador with Bitcoin Adoption https://finance.yahoo.com/news/world-bank-t-imf-might-112258013.html.

other SIDs face similar challenges to El Salvador. Seventy percent (70%) of the Jamaican population are unbanked and would benefit from remote financial intermediary solutions. The Bank of Jamaica was one of the first to establish its own digital currency JAMDEX which it plans to roll out in late 2022. The country recently started experimenting with a mobile money solution, NCB Lynk.

Along with digital money, the use of blockchain technology will make it easier for financial instruments and assets to move within and across borders. Compete Caribbean and the London School of Economics outline that blockchain has the potential to merge alternative information and reduce friction between transactions. This will allow these participants to increase their security on a platform that is accessible to everyone. Blockchain would substitute trust in humans with trust in mathematics and if done without entry error can be more accurate. Blockchain technology would reduce data vulnerability to manipulation and disjointed processes. The blockchain technology will assist firms to overcome problems associated with asymmetric information, collateral requirements, a lack of sufficient credit reporting agencies and internet data security and cybercrime. The blockchain technology is ideal for any process that deals with clearing and settling transactions. It significantly reduces the paperwork and bureaucracy, it provides security and automates post-trade transactions in transfer processes involving contract-based agreements and it is particularly useful in the transfer of a stock or any other asset or commodities. Importation and exportation of intermediate and final goods is often overcrowded with bureaucracy and paperwork that minimize the efficiency of these processes. The use of blockchain to check and balance processes, including lead time, departure and arrival time as well as instant notification of delays and changes to your order is increasing globally. Micro, Small and Medium Sized Enterprises (MSMEs) in developing countries can also use blockchain for their processes to minimize handicaps in tracking and clearing shipping to increase efficiency of their business processes. Overall, blockchain technology has the potential to help countries to improve their national productivity.

In Jamaica, there has been an increase in digital support services for traditional businesses. For example, COVID-19 saw the rise of food delivery services like Uber Eats like HUGO, QUICKPLATE and 7Krave from the private sector and THE ENDS from the government. Two years later, the aftermath of the biggest health pandemic in over 100 years continues to linger and dictate the pace of global economic activity.

Economists and policymakers are forced to think outside of the box to find plausible short-, medium- and long-term solutions to minimize the effects of such an economic downturn, build resilience and to restore macroeconomic progress as soon as possible. The beginning of a new mindset emerging in an unexpected shift in the global economic paradigm.

References

Ball, E. V., Wang, S. L., Nehring, R., & Mosheim, R. (2016). Productivity and economic growth in US agriculture: A new look. *Applied Economic Perspectives and Policy, 38*(1), 30–49.

Beugelsdijk, S., Klasing, M. J., & Milionis, P. (2018). Regional economic development in Europe: The role of total factor productivity. *Regional Studies, 52*(4), 461–476.

Chen, E. K. (1997). The total factor productivity debate: Determinants of economic growth in East Asia. *Asian-Pacific Economic Literature, 11*(1), 18–38.

Crespi, F., & Pianta, M. (2008). Demand and innovation in productivity growth. *International Review of Applied Economics, 22*(6), 655–672.

Daude, C., & Fernández-Arias, E. (2010). *On the role of productivity and factor accumulation in economic development in Latin America and the Caribbean.*

Felipe, J. (1999). Total factor productivity growth in East Asia: A critical survey. *The Journal of Development Studies, 35*(4), 1–41.

Geroski, P. A. (1989). Entry, innovation and productivity growth. *The Review of Economics and Statistics, 71*, 572–578.

Gomez-Salvador, R., Musso, A., Stocker, M., & Turunen, J. (2006). Labour productivity developments in the euro area (No. 53). *ECB Occasional Paper.*

Guillaumont, P. (2013). Assessing the economic vulnerability of small island developing states and the least developed countries. In *Understanding Small-Island Developing States* (pp. 20–46). Routledge.

Keynes, J. M. (1937). The general theory of employment. *The Quarterly Journal of Economics, 51*(2), 209–223.

Hsieh, C. T. (1999). Productivity growth and factor prices in East Asia. *American Economic Review, 89*(2), 133–138.

Lee, J. W., & Xuan, Y. (2019). Effects of technology and innovation management and total factor productivity on the economic growth of China. *The Journal of Asian Finance, Economics and Business, 6*(2), 63–73.

Myrdal, G. (1972). *Asian drama: An inquiry into the poverty of nations* (Vol. 1). Pantheon.

Ngo, M. N., & Nguyen, L. D. (2020). Economic growth, total factor productivity, and institution quality in low-middle income countries in Asia. *The Journal of Asian Finance, Economics, and Business, 7*(7), 251–260.

Timmer, M. P., Inklaar, R., O'Mahony, M., & Van Ark, B. (2011). Productivity and economic growth in Europe: A comparative industry perspective. *International Productivity Monitor, 21*, 3.

Timmer, M. P., & Los, B. (2005). Localized innovation and productivity growth in Asia: An intertemporal DEA approach. *Journal of Productivity Analysis, 23*(1), 47–64.

United Nations. (2017, May). *World economic situation and prospects: May 2017 Briefing. No. 102.* https://www.un.org/development/desa/dpad/publication/world-economic-situation-and-prospects-may-2017-briefing-no-102/

Uppenberg, K., & Strauss, H. (2010). *Innovation and productivity growth in the EU services sector*. Luxembourg: European Investment Bank.

Van Ark, B. (2016). The productivity paradox of the new digital economy. *International Productivity Monitor, 31*, 3.

Van der Eng, P. (2009). *Total factor productivity and economic growth in Indonesia* (The Australian National University Working Papers in Trade and Development, Working Paper [2009/01]).

CHAPTER 2

Total Factor Productivity and Labour Productivity Drivers in Jamaica

2.1 Background

Jamaica's inability to increase per capita income and to find a consistent steady state economic growth path is rooted in the decline of its Labour (LP) and Total Factor (TFP) Productivity both of which show declining trends between 1996 and 2019, see Fig. 2.1. In comparison, as outlined in Chapter 1, productivity in other Caribbean Countries, and other regions like North America, East Asia and Western Europe have all improved. Traditionally, knowledge accumulation and technological progress are two of the original factors that promote productivity growth among countries (Solow, 1965). However, as the structure of economies evolve and economies become more complexed, other factors emerge, some becoming more important while others, less important over the past 60 years (Dieppe et al., 2021).

The factors that influence labour productivity may or may not be the same that influence total factor productivity. Dua and Garg (2019) found evidence to suggest that labour productivity growth in Asia over the past half century can be attributed to policies that enabled capital deepening, higher human capital, technological progress, institutional quality, trade openness and reduction in government size. They analysed and compared the determinants of labour productivity between developed and developing countries of the Asian Pacific, while Loko and Diouf (2009) examined the main determinants of TFP growth in 62 countries using a

A. Haughton and W. Ivey, *Overcoming Productivity Challenges in Small Countries*, https://doi.org/10.1007/978-3-031-23301-2_2

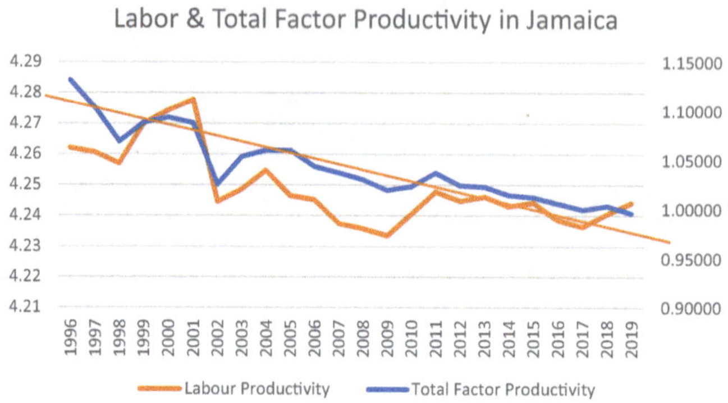

Fig. 2.1 Labor and total factor productivity in Jamaica

Principal Component Analysis (PCA) and a dynamic panel data model. Along with a case study that investigates reforms that might be necessary to boost gains from TFP in these countries, Loko and Diouf (2009) findings suggest that policies to attract foreign direct investment, reducing the size of the government, reallocating resources from low productivity sectors to higher ones and those that incentivize women to enter the workforce might improve TFP. The results also imply that reforms aimed at strengthening human capital, increasing the volume of trade, and improving the business environment can improve TFP. The 62 countries examined included a single Caribbean country Trinidad and Tobago. The remaining Caribbean countries especially Jamaica face productivity issues and would like to understand if the factors outlined above as drivers are relevant to them.

Figure 2.2 shows the opposing trajectory of TFP in Jamaica and its major trading partner, the United States for example. Cardarelli and Lusinyan (2015) suggest that TFP growth was a key component of US GDP growth spurts throughout the 90s and early 2000s, see Fig. 2.2.[1] North America currently has the highest income per unit of labour employed in the world. Labour productivity in the United States of America increased by 330%, from little over \$35,000 in 1950 to over \$114,000 in 2019. US per capita income of approximately \$65,000 is 10 times more than the average per capita income in Jamaica of just

[1] Notice the upward trend of TFP during this time.

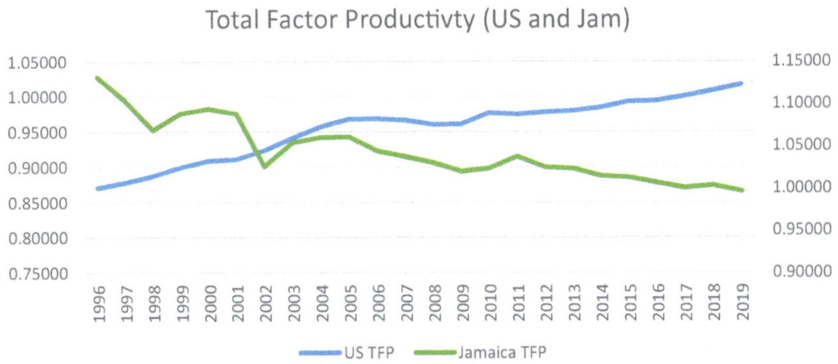

Fig. 2.2 The United States and Jamaica TFP (1996–2019)

$4,700. The clear decline in Jamaica's TFP is indicative that the country has not capitalized on the technological boom. Suggesting that other factors contribute to enhancing the impact of technology. Cardarelli and Lusinyan (2015) recommend that countries should increase the implementation of policies to promote the development of human capital and research and development. These will allow technological progress and technical efficiency to improve, pushing production levels closer to the country's production possibility frontier.

Historically, Jamaica's TFP and LP fell gradually between 1996 and 1998. During this time, Jamaica encountered a financial crisis (Haughton, 2017). Major Banks could not repay accumulated debt when the exchange rate depreciated severely after Jamaica liberalized its financial sector without sufficient controls. The government established the Financial Sector Adjustment Company (FINSAC) to coordinate the recovery and restructuring process. TFP and LP showed signs of improvement from the late nineties to early 2000s. TFP as well as LP declined in Jamaica between 2001 and 2002. TFP fell below that of the United States in 2002. IMF (2002) explained that the Jamaican economy suffered from a series of shocks in 2001; internally, from an outbreak of violence in July, and externally, from the 911 terrorist attack in the United States in September.[2] Rain and floods damaged infrastructure and agricultural

[2] For further details on Jamaica's economic performance during this time, see IMF country report, https://www.imf.org/external/pubs/ft/scr/2002/cr02197.pdf.

produce in late November. In response, the government implemented expansionary fiscal measures. There was a corresponding improvement in LP and TFP up to 2004. The fiscal expansion came at the expense of increase debt and Jamaica, according to the IMF almost plunged into another crisis. Jamaica accumulated a public debt of more than 140% of its GDP. The national debt had a negative effect on economic growth in Jamaica (Blavy, 2006).[3] Subsequent budgets focused on debt repayment and achieving primary surplus targets. Less focus was placed on infrastructural, institutional, structural or other spending that would be necessary to improve the nation's output and the value of each unit of labour employed. As a result, the nation's productivity challenges continued.

In 2005, Hurricane Ivan ravaged through the island, damaging crops and infrastructure. A ruthlessly active hurricane season mixed with drought reduced Jamaica's output. In 2006, Jamaica's economy showed signs of improvement. GDP grew by nearly 3%. Annual inflation of 6% was roughly 3.5% below the targeted 9.5% for the 2006/2007 fiscal year. Net International Reserves (NIR) reached historical highs. The improvement in Jamaica's economic condition was short-lived as Hurricane Dean inflicted more damages in 2007. These continuous events correspond with a decline in TFP and LP from 2004 up to 2009. TFP and LP displayed signs of improvement following the Global Financial Crisis, however, dampened not long after. Both reverted to the visible downward long-term trend from 2011 to 2017. Since then, TFP has continued to decline but LP increased marginally. Global productivity growth also slowed after the GFC and has further worsened with the COVID-19 pandemic which has disrupted trade and the flow of FDI (Dieppe et al., 2021). Irrespective of the marginal meanderings, the general trajectory is a declining slope which is undesirable.

Throughout the progression of its economy over the past two decades, Jamaica has failed at its productivity pursuits. The operative question is therefore, how can Jamaica change the trajectory of the productivity of its factor inputs? What are the main variables of interest? An investigation into the possible factors that contribute to increasing TFP and LP is important to understanding Jamaica's growth pursuits to provide lessons to other Caribbean islands and SIDs. The rest of the chapter is outlined as follows, Sect. 2.2 discusses the literature on labour as well as

[3] Blavy, R. (2006). *Public debt and productivity: The difficult quest for growth in Jamaica*.

total factor productivity. Section 2.3 discusses the possible variables that are considered as productivity drivers. Section 2.4 explores the data and methodological econometric model that is employed to explore productivity drivers in Jamaica. Section 2.5 discusses the results while Sect. 2.6 concludes and provide policy recommendations.

2.2 Literature

Samargandi (2018) employed a Fully Modified Ordinary Least Square (FMOLS) model to analyse the determinants of Labour Productivity in the Middle East and North African (MENA) countries. Their results suggest that the size of employment and compensation are negatively related to Labour productivity, while human capital and capital stock are positively related to labour productivity in these countries. In this chapter, we will employ a similar methodology to that of Samargandi (2018) to analyse the factors that impact LP and TFP in Jamaica. We will however employ other variables as suggested by other literature including Loko and Diouf (2009) and Dua and Garg (2019) to account for domestic idiosyncrasies in Jamaica. Van der Eng (2009), among others found evidence to suggest that policies to increase the level of human capital, reduce the countries' share in agriculture and the exploration of value-added industries could improve the value of labour employed. Additionally, they found that low and stable inflation, increased trade openness, higher FDI inflows, and better scores on the economic freedom index contributes to productivity growth.

Cardarelli and Lusinyan (2015) employed panel data models to analyse the factors or policies that might be necessary to enhance productivity across the United States. They employed a stochastic frontier analysis to determine if TFP growth across states is influenced by technological trends and how efficiently information technology is utilized. Here, technological progress would increase the productive capacity of the economy, shifting the Production Possibility Frontier (PPF) outwards thereby increasing the capacity of what the country can produce. Their results suggest that Information Technology production and the usage efficiency are insufficient to explain changes in productivity growth across states in the United States especially after early 2000s. Gordon (2013) who examined US output and productivity growth also found similar evidence. States like New Mexico and Oregon, with similarly high levels of Information Technology production and usage, demonstrate dissimilar

TFP growth statistics. If implementation of these information technologies was distributed inequitably across these states, then the result could justify the inefficiency in certain states and a worsening of the disparity among TFP growth rates. This along with the results of Cardarelli and Lusinyan (2015) highlight the idea that the level of Total Factor Productivity is not solely dependent on the technology or the technological boom, but also the economy's ability to absorb and employ this technology. Erken et al. (2009) found evidence to suggest that entrepreneurship, research and development, trade openness and labour force participation rate had an impact on productivity in 20 European Countries between 1970 and 2001.

Nachega and Fontaine (2006) investigates empirically the sources of aggregate output growth and the determinants of total factor productivity (TFP) in Niger between 1963 and 2003. A growth accounting analysis indicates that the erosion in output per capita over the sample period is due to the negative growth of both TFP and physical capital per capita. Sound macroeconomic policies, supported by official development assistance and structural reforms, are found to be key to raising TFP growth. These variables will be included in our analysis of total factor productivity and labour productivity. The issue at hand is about value creation. In other words, how can Jamaica increase the value of its output sustainably?

2.3 Productivity Drivers in the Jamaican Economy

The aim of this section is to analyse the factors that might contribute to Total Factor Productivity (TFP) and Labour productivity in Jamaica. Based on the literature explored above, there are many economic, structural, institutional and other variables that might influence the rate of productivity in a country. The relevant economic factors include inflation, government size, trade openness, human capital, capital stock and the level of Foreign Direct Investment (FDI). Structural and institutional factors include corruption control, government efficiency and agriculture as a share of GDP. Data on these variables for Jamaica between 1990 and 2019 are listed in Table 2.1 along with their sources and descriptions. Different research uses different techniques but agrees on the inclusion of these economic and structural factors. It is important to note that due to data availability constraints, there was no variable to measure the level of technology for Jamaica. Nevertheless, the effect of the other variables are analysed.

Table 2.1 List of variables

Variable	*Measure/definition*	*Data Source*
Labour productivity	GDP per employed persons	Penn World Table 9.1
Labour productivity	Output per worker	Federal Reserve of St. Louis
Total factor productivity	Total factor productivity index	Penn World Table 9.1
Inflation rate	Inflation, consumer prices (annual %)	World Development Indicators 2020
FDI	Foreign direct investment, net inflows (% of GDP)	World Development Indicators 2020
Trade openness	Trade (% of GDP)	World Development Indicators 2020
Financial openness	Domestic credit to private sector (% of GDP)	World Development Indicators 2020
Structural shift	Agriculture, value added (% of GDP)	World Development Indicators 2020
Human Capital	Human capital index, based on years of schooling and returns to education	Penn World Table 9.1
Capital stock	Capital stock at constant 2011 national prices (in mil. 2011US$)	Penn World Table 9.1
Corruption control	Corruption control index	The Conference Board
Government efficiency	Government efficiency Index	The Conference Board

2.3.1 *Inflation*

The Bank of Jamaica in 2017 switched to inflation targeting. Jamaica's inflation rate has fallen from more than 25% in 1996 to less than 10% in 2019, see Fig. 2.3. What effect has this decline in inflation had on Jamaica's productivity? The literature suggests that inflation is an indicator of the level of stability of the economy and should be included in the productivity analysis; Nachega and Fontaine (2006), Loko and Diouf (2009),[4] Cardarelli and Lusinyan (2015), Van der Eng (2009), and Dua and Garg (2019).

[4] Study on total factor productivity growth.

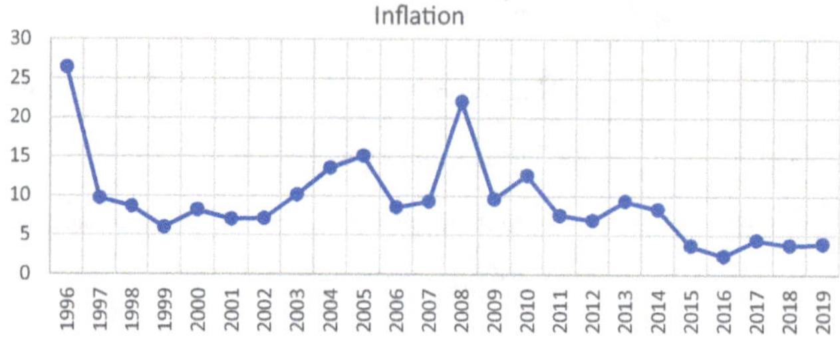

Fig. 2.3 Inflation

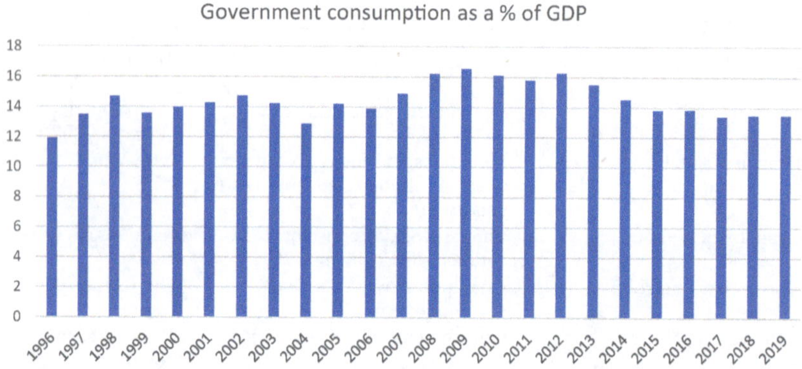

Fig. 2.4 Government consumption

2.3.2 *Government Size*

Government consumption as a percentage of GDP has fluctuated between 12 and 16.54% between 1996 and 2019, see Fig. 2.4. Since 2012, there has been a constant decline. The literature suggests that Government consumption as a percentage of GDP is an indicator to measure any structural shift. Cardarelli and Lusinyan (2015), Erken et al. (2009), Nachega and Fontaine (2006), Asghari and Heidari (2016), and Dar and AmirKhalkhali (2002), Loko and Diouf (2009).

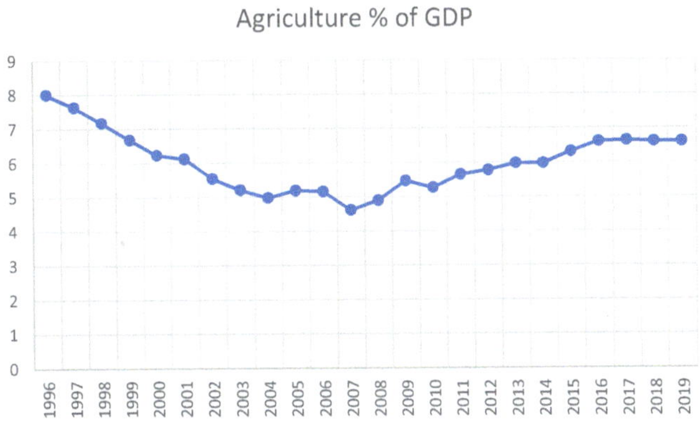

Fig. 2.5 Share of agriculture % GDP

2.3.3 *Share of Value Added in Agriculture*

Starting at 8%, Agriculture as a share of GDP in Jamaica gradually declined between 1996 and 2007 to about 4.6%. After which it gradually increased to 6.6% by 2019, see Fig. 2.5. There are more than 160,000 registered farmers in Jamaica alone, roughly 30% are female. Dua and Garg (2019), Loko and Diouf (2009), and Jaumotte and Spatafora (2007) found evidence to suggest that reducing the share of labour in the agricultural sector to industrial sector contribute to productivity.

2.3.4 *Human Capital*

Human capital was the single most important variable in driving productivity.[5] Loko and Diouf (2009), Cardarelli and Lusinyan (2015), Erken et al. (2009), Van der Eng (2009), Dua and Garg (2019), Popoola et al. (2019), found evidence to suggest that human capital is essential to labour productivity. After increasing sharply between 1996 and 2001, Jamaica's Human capital gradually decline up to 2010 and gradually increased up to 2019, see Fig. 2.6.

[5] Human Capital is constructed as an index based on average years of schooling by Barro and Lee (2013) and an assumed rate of return to education attainment which comes from the Mincer equation according to Psacharopoulos (1994).

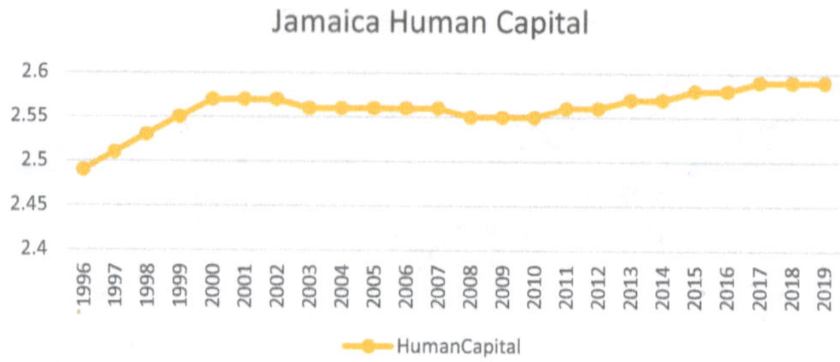

Fig. 2.6 Jamaica's human capital

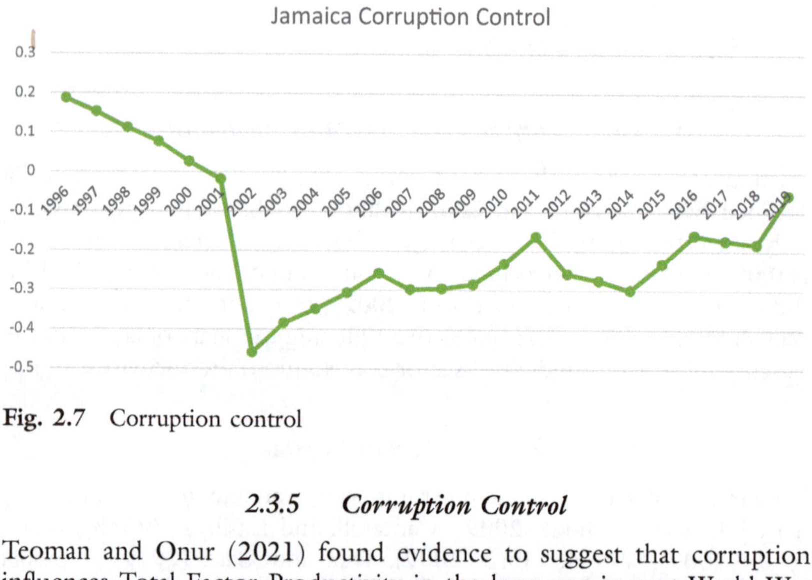

Fig. 2.7 Corruption control

2.3.5 *Corruption Control*

Teoman and Onur (2021) found evidence to suggest that corruption influences Total Factor Productivity in the long run in post-World War II in Turkey. Corruption has been listed as one of the hindrances to doing business in Jamaica.[6] Jamaica ranks high on the Global Corruption Index relative to other countries in the Caribbean as a result including the corruption variable is relevant. See for example Haughton (2017). Corruption control has been up and down over the last 20 years, see Fig. 2.7.

[6] National Five-Year Manufacturing Growth Strategy for Jamaica.

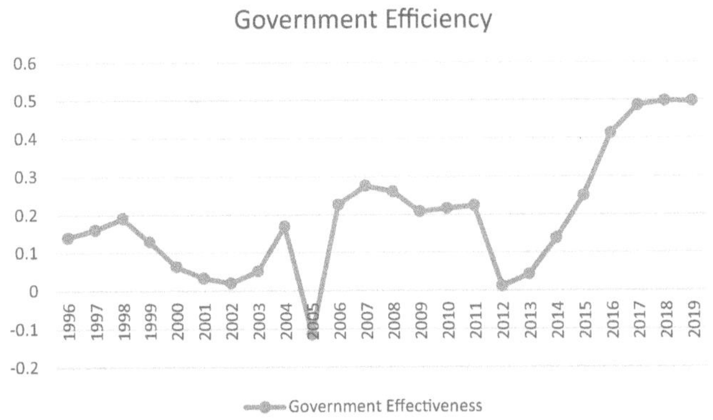

Fig. 2.8 Government of Jamaica efficiency

2.3.6 *Government Effectiveness*

Productivity in Jamaica is also dependent on Government Effectiveness. Government effectiveness shows a gradual increase since 2012 but demonstrates declining marginal increases from 2016 onwards, see Fig. 2.8.

2.3.7 *Capital Stock*

Figure 2.9 illustrates Jamaica's Capital stock from 1996 to 2019. Jamaica's Capital Stock is measured in millions of US dollars. For analysis purposes, we use a logged variable.

2.3.8 *Credit to the Private Sector*

Credit to the private sector can help to direct productive activity in country. If the private sector lends to the productive sector, production can increase. Credit to the private sector show signs of gradual increase over the last twenty years from 12% in 2000 to over 40% by 2019, see Fig. 2.10.

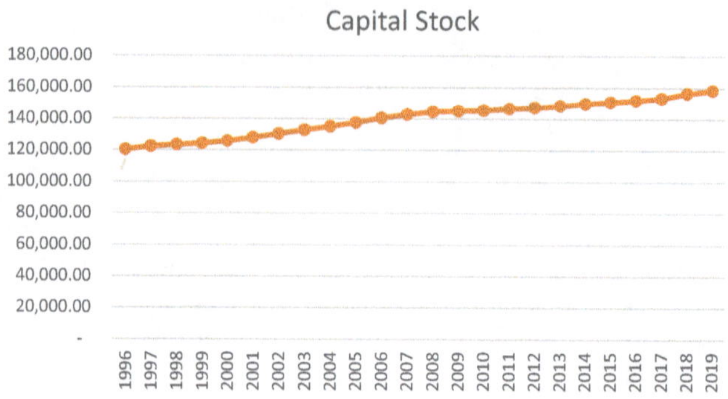

Fig. 2.9 Capital stock

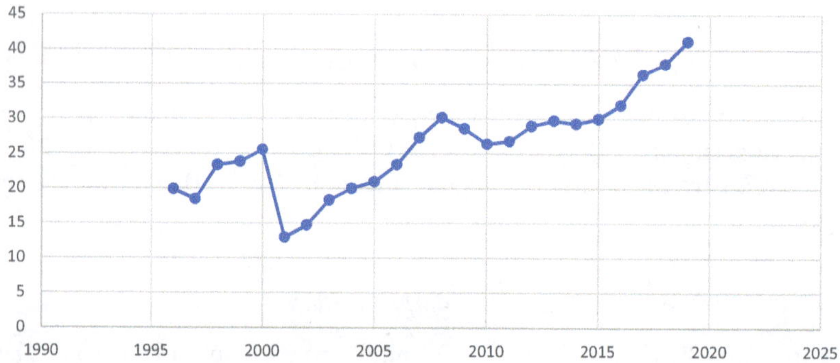

Fig. 2.10 Credit to the private sector by banks

2.3.9 *Trade Openness*

Dua and Garg (2019), Grossman and Helpman (1991), Sala-i-Martin and Barro (1995), Edwards (1997), Dar and Amirkhalkhali (2003), Freund and Bolaky (2008), Chang et al. (2009), Keho (2017), and Vlastou (2010) all found evidence to suggest that trade openness influences the level of growth of labour productivity in countries. From 2003 up to the Financial Crisis of 2008, Jamaica had become increasingly a more open economy trending from 81.9 to 114 in 2008. However, that trend was

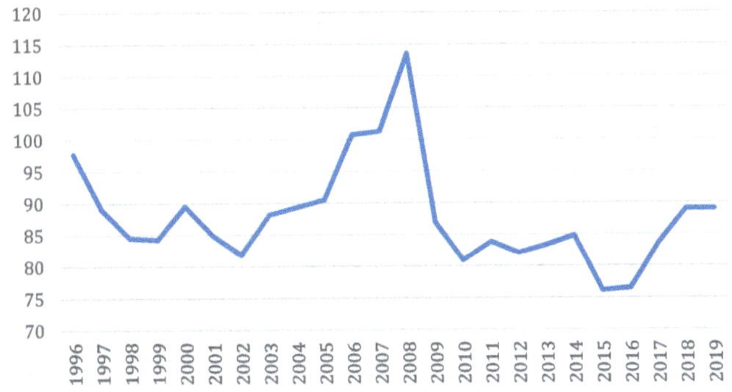

Fig. 2.11 Trade openness

reversed due to the hardships brought about by the crisis which saw a reduction in trade which gradually eroded the openness of the economy, see Fig. 2.11.

2.3.10 FDI

Loko and Diouf (2009) found evidence to suggest that FDI contributes to the level of labour productivity in the group of countries examined. While Haughton (2017) found evidence to suggest that FDI contribute negatively to GDP in Jamaica, it will be interesting to see the impact of FDI on productivity in Jamaica, see Fig. 2.12.

2.4 Data and Methodology

Jamaica's data is a specific case. Figure 2.1 shows that labour productivity and total factor productivity demonstrate similar downward meandering patterns and therefore might be impacted by similar variables outside of a few exceptions. Two separate equations are estimated to estimate the drivers of TFP as outlined in Eq. 2.1. The factors that might influence labour productivity are outlined by Eq. 2.2. Our analysis began in a general format but minimized to the specific models outlined in Eqs. 2.1 and 2.2 after a rigorous search and observe for significance process. The final factors that cointegrate with Jamaica's Total Factor Productivity

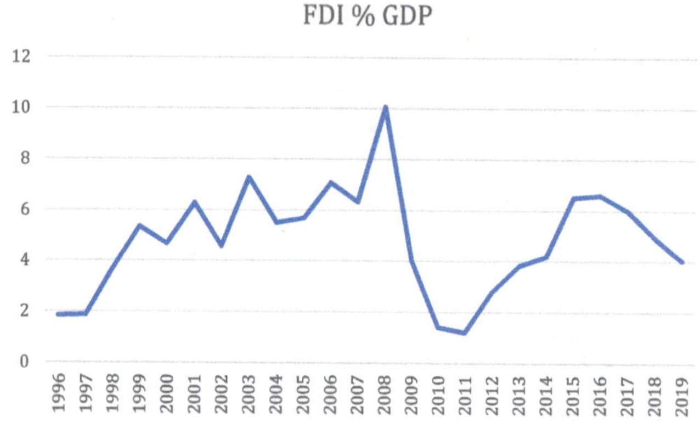

Fig. 2.12 FDI as a percentage of GDP

(TFP) and Labour Productivity (LP) are derived from Eqs. 2.1 and 2.2.

$$\begin{aligned}\text{TFP}_j =& \alpha + \varphi_1 y_t + \varphi_2 \text{inf}_t + \varphi_5 \text{GovtCon}_t + \varphi_6 \text{Crrupco} \\ &+ \varphi_7 \text{Agreva}_t + \varphi_9 \text{CPS}_t + \varphi_{10} \text{Capstock}_t + v_t \end{aligned} \tag{2.1}$$

$$\begin{aligned}\text{LP}_j =& \alpha + \varphi_1 y_t + \varphi_2 \text{inf}_t + \varphi_4 \text{HC}_t + \varphi_5 \text{GovtCon}_t + \varphi_6 \text{Inst}_t \\ &+ \varphi_7 \text{Struc}_t + \varphi_8 \text{Currupco}_t + \varphi_9 \text{CPS}_t \\ &+ \varphi_{10} \text{Capstock} + \varphi_{11} \text{femlf}_t + v_t \end{aligned} \tag{2.2}$$

TFP represents Total Factor Productivity, y_t is the initial value of GDP per capita, HC_t is human capital, Agriva is agriculture share of GDP, Govt_t is government size and or efficiency, FDI_t is foreign direct Investment, CPS_t is credit to the private sector, Struc is a set of structural factors including the EFI_t which is the economic freedom index, femlf_t is the female labour force and v_t is the common error term, respectively. Lagged dependent variables are employed to deal with any omitted variable problems, endogeneity or measurement errors that might arise. The empirical analyses are conducted in the following order:

1. Standard Augmented Dickie and Fuller (1979) and Phillips and Perron (1988) unit Root tests are conducted to verify the stationarity properties of each of the variables outlined as in Eqs. 2.1 and 2.2.

2. Upon ascertaining the stationarity properties of the variables, we test for a long-run relationship using the general to specific format (GETS) in the Fully Modified Ordinary Least Squares (FMOLS) estimation. To increase robustness, the results from the FMOLS are compared to the Canonical Cointegrating Regression (CCR).

2.4.1 Fully modified OLS

We adopt the method of Fully Modified Ordinary Least Squares by Phillips and Hansen (1990) which provides unbiased cointegrating estimators in a single equation model framework. The FMOLS accounts for endogeneity bias, auto or serial correlation, it is asymptotically unbiased and the most efficient estimator in its class. Define the following equation:

$$\ln P_t = \beta_0 + \beta_1 X_t + \mu_t \tag{2.3}$$

where P_t is LP_t in the labour productivity equation and TFP_t in the total factor productivity equation, X_t is a vector of regressors containing I(1) independent variables that are not cointegrated individually and μ_t is the error term. The difference stationary process of X_t can be expressed as

$$\Delta X_t = \varphi + \varepsilon_t \tag{2.4}$$

The FMOLS rectifies the long-run parameters in a two-step process. First, the dependent variable P_t is modified to be independent of μ_t and ε_t. The FMOLS the error vector $\psi_t = (\mu_t, \varepsilon_t)$ follows a strictly stationary process with mean zero and a finite positive covariance matrix Σ and the estimated residuals $\widehat{\mu_t}$ and $\widehat{\varepsilon_t}$ are independently and identically distributed.

Consider a $n+1$ time series vector process $\left(y_t, X_t^{'}\right)$ with a cointegrating equation as follows:

$$y_t = X_t' \boldsymbol{\beta} + D_{1t}' \xi_t + \mu_{1t} \tag{2.5}$$

where $D_t = \left(D_{1t}', D_{1t}'\right)'$ are deterministic trend regressors and the n stochastic regressors X_t

If μ_{1t} is the residuals obtained from estimating Eq. 2.5, then μ_{2t} can be obtained directly as $\mu_{2t} = \Delta\widehat{\varepsilon}_{2t}$.

$$X_t = \widehat{\Psi}_{21}' D_{1t} + \widehat{\Psi}_{22}' D_{2t}' + \widehat{\varepsilon}_{2t} \tag{2.6}$$

or by differencing the error obtained from the levels form in Eq. 2.7

$$X_t = \widehat{\Psi}'_{21} D_{1t} + \widehat{\Psi}'_{22} D'_{2t} + \widehat{\varepsilon}_{2t} \tag{2.7}$$

Let $\widehat{\Omega}$ and $\widehat{\Lambda}$ be the long-run covariance matrices computed using the residuals.

Where

$$\Sigma = E\left(\mu_t \mu_t'\right) = \begin{bmatrix} \sigma_{11} & \sigma_{12} \\ \sigma_{21} & \sigma_{22} \end{bmatrix}$$

$$\Lambda = \sum_{j=0}^{\infty} E\left(\mu_t \mu_{t-j}'\right) = \begin{bmatrix} \vartheta_{11} & \vartheta_{12} \\ \vartheta_{21} & \vartheta_{22} \end{bmatrix}$$

$$\Omega = \Lambda + \Lambda' - \sum = \sum_{j=-\infty}^{\infty} E\left(\mu_t \mu_{t-j}'\right) = \begin{bmatrix} \omega_{11} & \omega_{12} \\ \omega_{21} & \omega_{22} \end{bmatrix}$$

And the residual vector of innovations given by

$$\widehat{u}_t = \begin{pmatrix} \widehat{\mu}_{1t} \\ \cdot \\ \widehat{\mu}_{2t} \end{pmatrix} \text{ for } t = 2, 3, \ldots n$$

$\widehat{u_t}$ is strictly stationary and ergodic with zero mean, and contemporaneous covariance matrix Σ, and covariance matrix Ω each of which are partitioned comfortably with u_t. The matrix Ω is assumed to have a rank n with a nonsingular matrix Ω_{22}. Then the modified data can be defined as

$$y_t^+ = y_t - \widehat{\omega}_{12} + \widehat{\Omega}_{22}^{-1} \widehat{\mu}_2 \tag{2.8}$$

and the estimated bias correction term

$$\widehat{\vartheta}_t^+ = \vartheta_{12} - \widehat{\omega}_{12} + \widehat{\Omega}_{22}^{-1} \widehat{\Lambda}_{22} \tag{2.9}$$

And the FMOLS estimator is given by

$$\widehat{\zeta} = \begin{bmatrix} \widehat{\beta} \\ \cdot \\ \widehat{\xi}_1 \end{bmatrix} = \left(\sum_{t=2}^{T} Z_t Z_t^1\right)^{-1} \left(\sum_{t=2}^{T} Z_t y_t^+ - T \begin{bmatrix} \widehat{\vartheta}_{12}^{+'} \\ \cdot \\ 0 \end{bmatrix}\right) \tag{2.10}$$

where $Z_t = \left(X_t', D_t'\right)'$.

Table 2.2 Unit root test and order of integration for Jamaica

Variable	*Dickie Fuller*	*Phillips Peron*
TFP	I(1)	I(1)
LP	I(1)	I(1)
CPS	I(1)	I(1)
INFLATION	I(0)	I(0)
LCAPSTOC	I(1)	I(1)
CORUPCTL	I(1)	I(1)
GOVEFFECTIVE	I(1)	I(1)
GOVEMENTCONSUMP	I(1)	I(1)
AGRIVA	I(1)	I(1)
HC	I(1)	I(0)

Note Decision rule based on *P*-values $^{*}p<0.10$, $^{**}p<0.05$, $^{***}p<0.01$ indicate significance at the 10%, 5% and 1% level of significance, respectively

2.5 Results and Analysis

2.5.1 *Unit Root Test and Cointegration Test*

The results from the Augmented Dickey and Fuller (1979) and Phillips and Perron (1988) confirm that all of the variables except inflation are integrated of the order 1, or I(1) nonstationary in levels, see Table 2.2. Labour Productivity, Total Factor Productivity, credit to the private sector, inflation, capital stock, corruption control, government effectiveness, government consumption and agriculture as a share of GDP are all I(1). Given that the main variables of interest are I(1), cointegration tests can be applied. Two equations are estimated for TFP and LP coinciding with Eqs. 2.1 and 2.2.[7]

2.5.1.1 Total Factor Productivity

TFP represents the combination of labour and capital to produce output which depends on the right infrastructure as well as other structural and institutional factors that should be in place to minimize bureaucracy and other hindrances to efficient doing business and production processes. The results for the TFP cointegrating equation are shown in Table 2.3. The Fully Modified OLS and the Canonical Cointegration Regressions

[7] All the original variables were included in both regressions and we moved from the general to specific based on significance level and goodness of fit.

show similar results. The value of the constant in the regression is roughly 3.16, illustrating that Total Factor Productivity in Jamaica would be high if it were not for the negative effects of most of the dependent variables employed. TFP in Jamaica is negatively related to credit to the private sector, local capital stock, government effectiveness, government consumption and agriculture share of GDP. TFP positively related to inflation and corruption control. The inflation results are consistent with the theory on the relationship between the Phillips Curve and Okun's law. Inflation is positively related to total factor productivity. Contrary to the literature, see for example Loko and Diouf (2009), lower inflation makes the country more stable and as a result, more potential to grow. It appears that low domestic demand in these countries is burdened by high supply. One creating inflationary pressure while the other creating deflationary pressure. About a decade ago, Jamaica was a high interest rate regime. More recently, the government started practising inflation targeting which has been instrumental in lowering inflation but output has not been increasing.

As it relates to corruption, a 1% increase in Jamaica's efforts to control corruption can increase Total Factor Productivity by 0.14 or 14% on the index. Over the years, Jamaica has been criticized for its level of corruption. The Global Doing Business Reports often highlight the importance

Table 2.3 TFP model results

TFP	*Fully Modified Least Squares (FMOLS)*		*Canonical Cointegrating Regression (CCR)*	
	Coefficient	*P-value*	*Coefficient*	*P-value*
CPS	−0.000749	0.0947	−0.001091	0.1138
LCAPSTOCK	−0.157729	0.0012***	−0.138877	0.0364**
CORUPCTL	0.137547	0.0000***	0.143996	0.0000***
GOVEFFECTIVE	−0.061389	0.0001***	−0.060576	0.0006***
GOVCONSUMP	−0.006883	0.0001***	−0.006624	0.0001***
AGRIVA	−0.016445	0.0000***	−0.016103	0.0001***
C	3.164342	0.0000***	2.944839	0.0006***
	R-squared	0.951072	R-squared	0.944443
	Adjusted R-squared	0.932724	Adjusted R-squared	0.923609
	S.E. of regression	0.008254	S.E. of regression	0.008795
	Long-run variance	2.76E-05	Long-run variance	2.76E-05

Table 2.4 TFP Hansen parameter instability cointegration test

Hansen parameter instability cointegration test					
		Stochastic	*Deterministic*	*Excluded*	
TFP	Lc statistic	Trends (m)	Trends (k)	Trends (p2)	Prob.*
FMOLS	1.325741	6	0	0	<0.01
CCR	0.531933	6	0	0	>0.2*
Null hypothesis: Series are cointegrated					

of increasing corruption control in Jamaica. Extortion, nepotism in the offering of government contracts, white-collar collusion among other factors all contribute to the level of corruption in Jamaica. Chapter 8 takes an in-depth look at the corruption literature and the economic consequences of corruption in Jamaica. The results of the Hansen stabilization test for the FMOLS and the Canonical Cointegrating Regression show that the null hypothesis of cointegration cannot be rejected, as a result, the variables employed impact TFP in the long run, see Table 2.4.

Government consumption and government effectiveness impact TFP negatively in Jamaica. The literature agrees that higher government consumption reduces TFP but not higher government efficiency. Unlike previous studies, capital deepening, improvements in institutional quality and trade openness are not as relevant in the case of Jamaica. Overall, the model appears to be a good fit. Five of the seven variables included are significant at the 1% level while credit to the private sector is significant at the 5% level and capital stock is significant at 10%. The cointegrating equation can explain 94% of the variation in TFP, with R-squared of 0.96 and adjusted r-squared over 0.94 level of significance.

2.5.1.2 Labour Productivity

The results for the LP cointegrating equation are illustrated in Table 2.5. LP in Jamaica is negatively related to credit to the private sector, capital stock, government effectiveness and Rule of law, while positively related to Human Capital and corruption control. Initial value of roughly 3.89 on the index, a 1 unit increase in human capital causes a 0.298 increase in Labour productivity. A 1 unit increase in Jamaica's efforts to control corruption can increase Labour Productivity by 0.07 on the index. The literature agrees that higher government consumption reduces efficiency of labour. Overall, the model appears to be a good fit. Six of the

Table 2.5 LP Model Results

LP	*Fully Modified Least Squares (FMOLS)*		*Canonical Cointegrating Regression (CCR)*	
	Coefficient	P-*value*	*Coefficient*	P-*value*
CPS	−0.000509	0.0095***	−0.000493	0.1324
LCAPSTOCK	−0.035533	0.0515*	−0.034606	0.2341
CORUPCTL	0.066352	0.0000***	0.067166	0.0000***
GOVEFFECTIVE	−0.026408	0.0001***	−0.028255	0.0007***
HC	0.342915	0.0000***	0.338007	0.0000***
DUMMY1	0.011292	0.0003***	0.009515	0.0644*
RLAW	−0.019504	0.0080***	−0.018357	0.0602*
C	3.815121	0.0000***	3.817314	0.0000***
	R-squared	0.960021	R-squared	0.956593
	Adjusted R-squared	0.941365	Adjusted R-squared	0.936336
	S.E. of regression	0.002926	S.E. of regression	0.003049
	Long-run variance	4.58E-06	Long-run variance	4.58E-06

seven variables included are significant at the 1% level while the capital stock is significant at the 10%. The cointegrating equation can explain 94% of the variation in LP, with an R-squared of 0.96 and adjusted r-squared over 0.94 level of significance. Inflation, government consumption, Foreign Direct Investment, trade-openness and financial deepening have no significant bearing on labour productivity in Jamaica. The Canonical Cointegrating Regression supports the result of the FMOLS, see Table 2.5. The results of the Hansen stabilization test for the FMOLS and the Canonical Cointegrating Regression show that the null hypothesis of cointegration cannot be rejected, as a result, the variables employed impact LP in the long run (see Table 2.6).

2.6 Conclusion and Recommendations

Labour as well as total factor productivity have been on the decline over the past two decades in Jamaica. Among all the countries in the Caribbean, Jamaica has had the lowest level of labour productivity over the past two decades. This chapter employs the Fully Modified OLS to investigate if there is any long-run relationship between LP or TFP and dependent variables suggested by the literature. The results are compared

Table 2.6 LP Hansen parameter instability cointegration test

Hansen parameter instability cointegration test					
		Stochastic	*Deterministic*	*Excluded*	
LLP	Lc statistic	Trends (m)	Trends (k)	Trends (p2)	Prob.*
FMOLS	0.686918	7	0	0	0.1527*
CCR	0.563624	7	0	0	>0.2*
Null hypothesis: Series are cointegrated					

to the output of the Canonical Cointegrating Regression for robustness. Both techniques are highly efficient and account for any endogeneity and serial correlation that might exist. The Hansen Stability test for cointegration is also employed for both TFP and LP for both cointegration methods. Annual data for TFP, LP and the dependent variables are collected from various sources from **1996** to **2019**. A graphical examination shows that labour productivity and total factor productivity move along similar trajectory, as a result, the factors that might contribute to total factor productivity are also investigated. The results suggest that policies to increase corruption control will contribute positively to Total Factor Productivity in Jamaica. The size of government, Financial Development, agricultural value added as a percentage of GDP and government effectiveness are negatively related to TFP. Unlike previous studies, capital deepening, improvements in institutional quality and trade openness are not as relevant in the case of Jamaica. The results suggest that policies to improve the level of human capital as well as increase corruption control will contribute positively to Labour Productivity in Jamaica. Government consumption is negatively related to labour productivity. Inflation, government consumption, Foreign Direct Investment, trade openness and financial deepening have no significant bearing on labour productivity in Jamaica. Here, we recommend that the government of Jamaica focus on expanding the human capital capacity of the country, improve corruption control and reduce the agricultural share of GDP. Policies to sophisticate the domestic economy similar to Trinidad who have diversified into the production of oil and natural gas are also important for Jamaica. The country should also diversify away from primary to secondary to tertiary production of goods similar to the Dominican Republic who have seen increased productivity over similar time frame.

References

Asghari, R., & Heidari, H. (2016). An investigation of the impact of size of the government on economic growth: Some new evidence from OECD-NEA Countries. *Iranian Economic Review, 20*(1), 49–68.

Barro, R. J., & Lee, J. W. (2013). A new data set of educational attainment in the world, 1950–2010. *Journal of Development Economics, 104*, 184–198.

Blavy, R. (2006). *Public debt and productivity: The difficult quest for growth in Jamaica*.

Sala-i-Martin, X. X., & Barro, R. J. (1995). *Technological diffusion, convergence, and growth* (No. 735). Center Discussion Paper.

Cardarelli, M. R., & Lusinyan, M. L. (2015). *US total factor productivity slowdown: evidence from the US states*. International Monetary Fund.

Chang, R., Kaltani, L., & Loayza, N. V. (2009). Openness can be good for growth: The role of policy complementarities. *Journal of Development Economics, 90*(1), 33–49.

Dar, A., & Amirkhalkhali, S. (2003). On the impact of trade openness on growth: Further evidence from OECD countries. *Applied Economics, 35*(16), 1761–1766.

Dar, A. A., & AmirKhalkhali, S. (2002). Government size, factor accumulation, and economic growth: evidence from OECD countries. *Journal of Policy Modeling, 24*(7–8), 679–692.

Dickey, D. A., & Fuller, W. A. (1979). Distribution of the estimators for autoregressive time series with a unit root. *Journal of the American Statistical Association, 74*(366a), 427–431.

Dieppe, A., Francis, N., & Kindberg-Hanlon, G. (2021). Technological and non-technological drivers of productivity dynamics in developed and emerging market economies. *Journal of Economic Dynamics and Control, 131*, 104216.

Dua, P., & Garg, N. K. (2019). Determinants of labour productivity: Comparison between developing and developed countries of Asia-Pacific. *Pacific Economic Review, 24*(5), 686–704.

Edwards, S. (1997). Trade policy, growth, and income distribution. *The American Economic Review, 87*(2), 205–210.

Erken, H., Donselaar, P., & Thurik, R. (2009). *Total factor productivity and the role of entrepreneurship*. Tinbergen Institute Discussion Paper.

Freund, C., & Bolaky, B. (2008). Trade, regulations, and income. *Journal of Development Economics, 87*(2), 309–321.

Gordon, R. J. (2013). US productivity Growth: The Slowdown has returned after a temporary revival. *International Productivity Monitor, 25*, 13.

Grossman, G. M., & Helpman, E. (1991). Trade, knowledge spillovers, and growth. *European Economic Review, 35*(2–3), 517–526.

Haughton, A. (2017). Balance of payments and capital account sustainability. In *Developing sustainable balance of payments in small countries* (pp. 89–102). Palgrave Macmillan.

Jaumotte, F., & Spatafora, N. (2007). *Asia rising: A sectoral perspective*. Available at SSRN 995630.

Keho, Y. (2017). The impact of trade openness on economic growth: The case of Cote d'Ivoire. *Cogent Economics & Finance, 5*(1), 1332820.

Loko, M. B., & Diouf, M. A. (2009). *Revisiting the determinants of productivity growth-what's new?* International Monetary Fund.

Nachega, J. C., & Fontaine, T. (2006). *Economic growth and total factor productivity in Niger*.

Popoola, O., Alege, P. O., Gershon, O., & Asaleye, J. A. (2019). Human capital channels and productivity growth: Evidence from Nigeria. *Economics and Sociology, 12*(4), 59–73.

Psacharopoulos, G. (1994). Returns to investment in education: A global update. *World Development, 22*(9), 1325–1343.

Phillips, P. C., & Hansen, B. E. (1990). Statistical inference in instrumental variables regression with I (1) processes. *The Review of Economic Studies, 57*(1), 99–125.

Phillips, P. C., & Perron, P. (1988). Testing for a unit root in time series regression. *Biometrika, 75*(2), 335–346.

Samargandi, N. (2018). Determinants of labor productivity in MENA countries. *Emerging Markets Finance and Trade, 54*(5), 1063–1081.

Solow, R. M. (1965). Technology and unemployment. *The Public Interest, 1*, 17.

Teoman, Ö., & Onur, Y. E. N. İ. (2021). Evaluation of the cluster components for apiculture products and honey in the black sea region of Turkey. *Eskişehir Osmangazi Üniversitesi İktisadi ve İdari Bilimler Dergisi, 16*(1), 34–54.

Van der Eng, P. (2009). *Total factor productivity and economic growth in Indonesia*. The Australian National University Working Papers in Trade and Development, Working Paper (2009/01).

Vlastou, I. (2010). Forcing Africa to open up to trade: Is it worth it? *The Journal of Developing Areas, 44*, 25–39.

CHAPTER 3

Productivity and Labour Usage in the Goods Producing Sector in Jamaica

Economic growth in any country is highly dependent on how efficiently labour is employed by firms in each industry. The more efficiently labour is employed the more output the firm can produce per person and the higher the nation's output growth. On the other hand, if labour is inefficiently employed, the more people the firm will have to hire to increase output through the extensive rather than the intensive margin, which is undesirable. There are instances when hiring more people reduces the overall productivity and output of the firm like a "too many cooks spoil the broth" scenario. This has been a longstanding issue in Jamaica. Carter (1997) noted that in Jamaica, a significant portion of the workforce could be laid off without any commensurate reduction in output.[1] Carter (1997) also noted that this observation is valid whether the groups consist of managers, supervisors or rank and file workers in the private or public sector. The exact number of workers that could be laid off from each industry is unknown, and therefore, the aim of this chapter is to estimate this for the goods producing industries in Jamaica while Chapter 4 does the same for the service producing industries.[2]

[1] Carter. (1997). *Why workers won't work: The worker in a developing economy: A case study of Jamaica*. https://books.google.com.jm/books?id=6hNGAAAAYAAJ&q=why+workers+wont+work&dq=why+workers+wont+work&hl=en&sa=X&redir_esc=y.

[2] Goods Producing Industries are all those industries that make a tangible commodity. For example, Agriculture, Manufacturing and Mining.

A. Haughton and W. Ivey, *Overcoming Productivity Challenges in Small Countries*, https://doi.org/10.1007/978-3-031-23301-2_3

According to the marginal productivity and bargaining theories, higher output per worker is an incentive for profit-seeking firms to hire more people, who as a result should receive more compensation (Yildirim, 2015). The opposite should also be true. Declining output per worker should push profit-seeking firms to hire less people as real wages fall. This however, is not the case in Jamaica where increasing employment is associated with declining labour productivity and falling real wages, see Fig. 3.1. The situation is similar to what happened in the OECD between 2014 and 2019, where the rise in employment was driven mainly by low productivity and low wage jobs. However, different from the United States, where increasing labour productivity has been associated with a fall in employment in some sectors but an increase in employment overall (Nordhaus, 2005). According to Landmann (2005), employment and productivity should be procyclical, normally move together throughout business cycles. When employment is increasing, labour productivity should be increasing and vice versa. Romer (1999) suggests that of the nine recessions that the United States encountered after the post-war era, employment fell in all nine and labour productivity fell in eight of the nine. But in Jamaica, employment is anti-cyclical with productivity and real wages. In essence, productivity decreases as the employed labour force gets larger, see Fig. 3.2.

Many developing countries hire people as a means of poverty alleviation rather than to increase production. South Africa is one such example in recent years, Jamaica seems to be another. In many of these developing countries, there are many people employed to redundant or low productivity jobs that do not sufficiently contribute to the total value added of the firm in those industries. Their compensation and therefore, their standard of living, which are derivatives of the income from their output, continue to decline. For example, in 2019, Jamaica's unemployment rate fell by 5.5% percentage points, from about 12.7% in 2017 to 7.2%. More people employed should translate into more output based on Okun's law but Jamaica continued to grow at its 50-year average of 1% per annum or a cumulative 2% for the two years.

By examining the elasticity of output to unemployment changes, the coefficient is less than 1 in absolute value indicating therefore that output in Jamaica is inelastic to reductions in unemployment.[3] Okun's law

[3] It is necessary to calculate a simple two-period output elasticity to unemployment $\%\Delta\text{Output}/\%\Delta\text{unemployment} = 2/5.5 = 0.36$, which is different from the time series

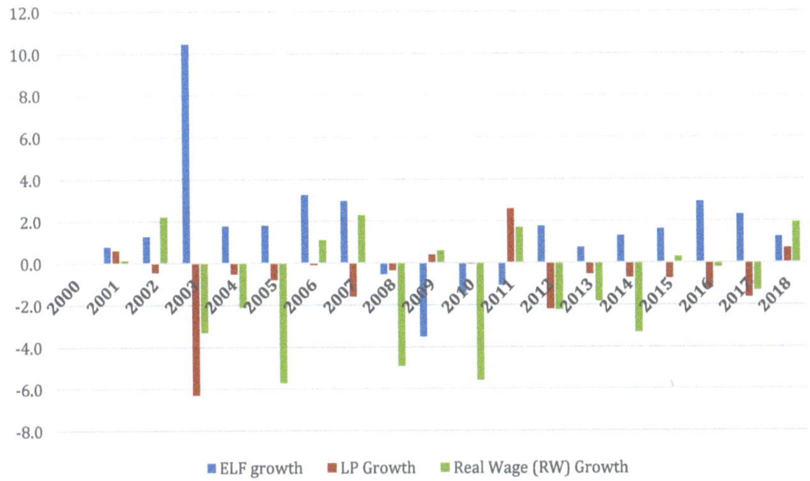

Fig. 3.1 Labour productivity, real wage and employment growth (*Source* STATIN)

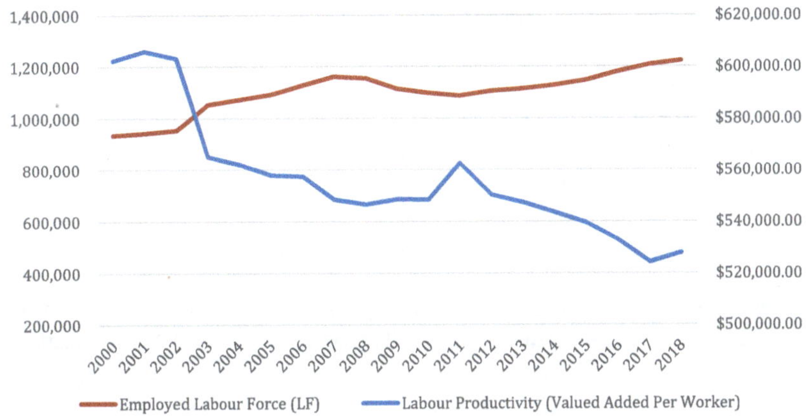

Fig. 3.2 Labour productivity and employed labour force (*Source* STATIN)

or Cob Douglass approach as outlined in the literature, see for example https://papers.ssrn.com/sol3/papers.cfm?abstract_id=1240656, https://www.jstor.org/stable/4190378?seq=3#metadata_info_tab_contents.

evidence in the United States shows a 3:1 or 2:1 ratio of output to unemployment, i.e. output increases by 3 or 2% to a 1% fall in unemployment (Mankiw, 2010). In the case of the United States, output is elastic to employment changes given that the percentage change in output is greater than the percentage change in unemployment. An inelastic output to employment ratio in Jamaica means that Jamaica can benefit from understanding its labour to output dynamics in each sector. It is essential to ascertain the optimal level of employment needed to yield maximum output in order to minimize the underutilization of labour employed. Fundamentally, labour employed ineffectively signals a waste of resources.

Many of these underutilized persons could be better employed elsewhere. By finding the level of labour where productivity is highest among each person employed, the excess labour can be filtered into other productive use. Sectorial analysis and productivity planning is crucial if any country is to achieve their developmental pursuits (UN World Development Survey). The low level of output per person and low associated wages are pervasive in all sectors but is more severe in the goods producing industries than the service producing industries in Jamaica. Figure 3.3 shows labour productivity by industry in Jamaica in 2018. Of all the industries, Agriculture Forestry and Fishing is the least productive with output per worker of just $1,860 USD per annum. Electricity, Gas and Water is the most productive labour of $24,289 USD per annum. The low level of output per worker in Jamaica is a cause for concern and requires an investigation into the labour usage mechanics of each industry if the country is to improve.

3.1 The Optimal Level of Employment by Industry

In total, there are four goods producing industries in Jamaica; (1) Agriculture, Forestry and Fishing (AFF), (2) Construction, (3) Manufacturing and (4) Mining and Quarrying. Output in the first three have remained relatively low and flat compared to Mining and Quarrying over the last 20 years, which has declined in a meandering manner, see Fig. 3.4. In terms of productivity growth, the Agriculture, Forestry and Fishing industry has seen a slight growth in average productivity of 1% per annum since 2002. All the other goods producing sectors' labour productivity has declined on average over the last 20 years. Mining and Quarrying has declined on average by 2.44% per annum or by 30% in total over the last 20 years. Manufacturing has declined by 1.1% on average or 17% in total

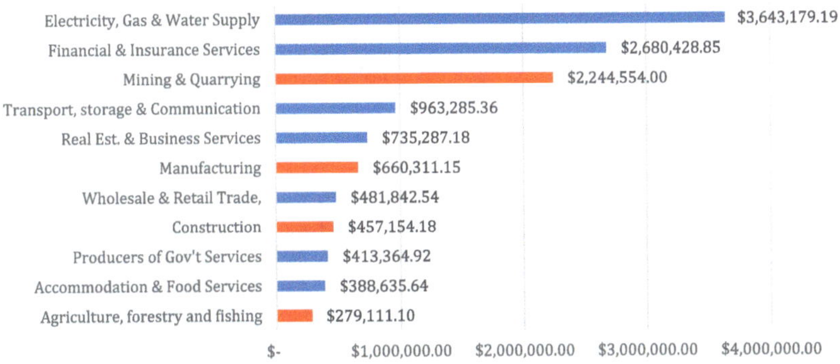

Fig. 3.3 Output per worker by industry 2018 (*Source* STATIN)

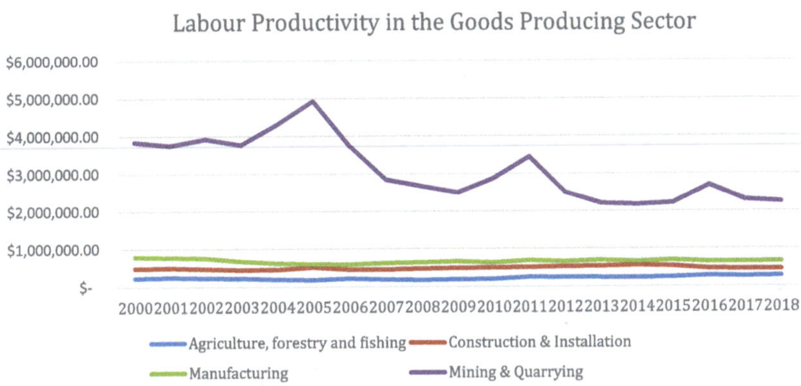

Fig. 3.4 Productivity growth in the goods sector (*Source* STATIN)

and Construction has declined by 0.13% on average or 2.3% in total, see Table 3.1.

The aim is to ascertain the optimal amount of labour that Jamaica must hire in each of these goods producing industries to be most efficient and not employ more people than what is needed. For each industry, a maximum output threshold is established using the least amount of labour that yields the greatest amount of output based on data collected over the last 20 years, thereby calculating the excess labour employed in each industry. In theory, if x amount of people produced y amount of output each year, then using x + 1 to produce the same amount of y

Table 3.1 Summary statistics

	Mean	*Median*	*Min*	*Max*	*Sum*	*Stand. Dev*	*Sample variance*	*Range*	*Standard error*
Agriculture	1.03	−1.74	−12.03	24.37	17.52	10.33	106.76	36.39	2.51
Mining & Q	−1.81	−2.44	−27.51	21.49	−30.72	15.47	239.45	49.00	3.75
Manufac.	−1.05	−2.06	−10.83	9.57	−17.85	5.73	32.89	20.39	1.39
Construc.	−0.13	1.36	−13.14	11.40	−2.29	6.14	37.69	24.54	1.49

or less is a waste of labour that could be used elsewhere, ceteris paribus. If the new person hired reduces total output or output per person, they can be considered as excess labour and can be removed from the labour force of that specific industry. The amount of labour that produces the most output per person is the labour maximizing level. Hiring labour above or below this leads to inefficiency, under current circumstances holding all other factors constant. The under-employed labour in these production process realizes less than optimal output.For visualization of the scenario, scatter plots are used to illustrate the pairwise maximization in each industry.[4]

The following are estimated; (1) The optimal ($L_{optimal}$) amount of labour employed (Eq. 3.1). (2) The excess labour (L_{excess}) in each sector (Eq. 3.2). (3) The optimal ($GDP_{optimal}$) amount of GDP that the country could produce if the currently employed labour were producing at ($L_{optimal}$), see Eq. 3.3. (4) The excess GDP (GDP_{excess}) that could be added if current labour in each industry was optimally employed in Eq. 3.4. (5) The optimal level of labour productivity ($LP_{optimal}$) for each person employed see, Eq. 3.5. (6) The excess labour productivity (LP_{excess}) that could be added if current labour was optimally emoplyed, see Eq. 3.6. LP_{MAX} represents the maximum level of productivity that could be reached if by employing ($L_{optimal}$).

$$L_{optimal} = f(L, LP_{MAX}) \tag{3.1}$$

$$L_{excess} = L_{optimal} - L_{current} \tag{3.2}$$

$$GDP_{optimal} = f\left(L_{optimal}, LP_{MAX}\right) \tag{3.3}$$

$$GDP_{excess} = GDP_{optimal} - GDP_{current} \tag{3.4}$$

$$LP_{optimal} = f\left(L_{optimal}\ ,\ LP_{MAX}\right) \tag{3.5}$$

$$LP_{excess} = LP_{optimal} - LP_{current} \tag{3.6}$$

[4] Different from Vo et al. (2019) who used daily indices for 10 industries for each country, there is no daily industry indices available for Jamaica and therefore annual labour productivity data is used.

3.2 Agriculture in Jamaica

Jamaica, once known as the pearl of the Antilles flourished from the productivity of its agricultural sector. Through specialization in mainly sugar cane and banana, Jamaica benefited from economies of scale. Earnings from the sector created a surplus of foreign exchange used to support Jamaica's Balance of Payments (BOP) and to purchase machinery and equipment to assist on its path to industrialization. As early as the 1930s, Kingston the capital of Jamaica, was on its way to modernization in logistics and infrastructure built on the back of the proceeds from agriculture. There were several advantages to having a productive agricultural sector; according to Lewis (1954) and Johnson (1997), an increase in agricultural output reduced the price of food, and therefore increases real wages and the standard of living of the citizens of a country. Lower cost of food also helps to keep real wage costs low for firms in the industrial sector thereby leaving excess resource to be invested in other capital, technology and infrastructure. According to Johnston and Mellor (1961), agriculture play 5 roles in the development process. These include (1) Increase the supply of domestic savings, (2) Earn foreign exchange, (3) Increase the supply of food for domestic consumption, (4) Release labour for industrial expansion and (5) Enlarge the size of the market for industrial output.

After the Second World War, emerging economic developmental models theorized a trade-off between agricultural expansion and industrialization. Clarke (1991) and Kuznets (1954) outlined the declining role of agriculture in the growth and development of modern economies. Lewis (1954) highlights that it would be necessary to relocate workers from the rural areas where they receive low wages from the practice of agriculture to the cities where they can earn higher wages from industrialization. According to Timmer (2002), the idea that a country's modernization and industrialization had to occur at the expense of a declining agricultural sector was widespread in the minds of many leaders in developing countries. As a result, the policies, investment and legislative framework that would have been necessary to allow farming to thrive were not initiated in many of these developing countries including Jamaica. Additionally, in Jamaica, sugar cane, banana and other main agricultural crops lost their markets and preferential treatments resulting in a decline in output. Sugar cane output declined from over roughly 5 million tonnes in 1968 to less than a million tonnes in 2019; more than 80% decline in production, see Fig. 3.5. Banana output also declined from more than

240,000 tonnes in 1968, to roughly 60,000 tonnes in 2019. A similar 80% decline like sugar cane, see Fig. 3.6. In general, agriculture has been declining in its contribution to economic growth ever since.

The shift in economic focus from agriculture to industrialization following the Second World War was real. The smaller the share of agriculture the more likely the rapid growth of the economy according to Timmer (2002) as well as Syrquin (1988). There was no push to incorporate technology, machinery and modern science to improve agricultural productivity. Despite the slowdown in large-scale production of traditional crops like sugarcane and banana, agriculture in Jamaica continued to survive as a traditional way of life and community-based subsistence practice. The large-scale farms by few owners have been replaced by many small farms with many owners. A good portion are very small plots of land sometimes with uneven hilly terrain. Jamaican agricultural goods are of better quality than the other countries in taste, active ingredients and potency but are limited in quantity, not enough to furnish a global demand. Agricultural goods rarely compete on quality, they compete in quantity and prices, with the bulk of goods coming from other countries with larger output of lesser quality. Recently, over the 20-year period (2000–2018) Jamaica agriculture industry output increased from $41

Fig. 3.5 Jamaica Sugarcane output

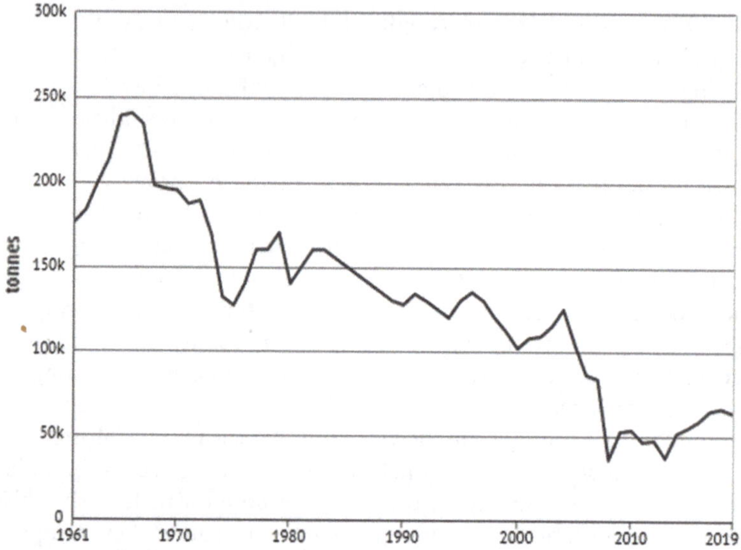

Fig. 3.6 Jamaica Banana output

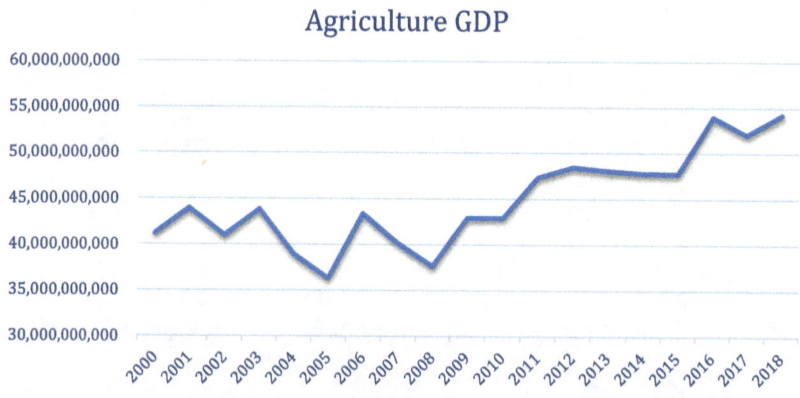

Fig. 3.7 Agriculture GDP (2000–2018)

billion to $54 billion, see Fig. 3.7. This represents a $13 billion JMD increase over 20 years or 0.7% growth per annum, but still not enough to compensate the large number of workers employed to the industry.

Over the last 20 years, the Agriculture, Forestry and Fishing industry has had the lowest value of output per person among all the industries in Jamaica. This translates to $260,000 JMD or $1,700 USD on average per person annually which is lower than both GDP per capita and the national minimum wage. Each person employed in this industry, grows, transports and sells roughly $22,000 JMD or $150 USD worth of goods per month and $5,500 JMD or **$36 US per week**, less than a dollar US an hour for a 40-hour workweek. The agricultural industry makes up about 7% of Jamaica's GDP and hires 17% of the country's employed labour force. Approximately 240,450 people work in the industry, the second largest employer of Jamaican people. The agricultural industry is highly concentrated with low-skilled labour earning less than $1 US per day which puts them below the poverty line. The government sends experienced farmers on farm work abroad. These farmers neglect their own plots of land to tender their trade elsewhere.

Figure 3.7 shows that agricultural GDP displays volatility over the last two decades. Agricultural productivity also displays a similar level of volatility up to 2019, see Fig. 3.8. According to Lucas (1973), instability in the agricultural sector has three effects on the economy. (1) Reduces the quality of investment by shifting expected returns. (2) Triggers volatility of prices in other sectors across the economy and (3) Increases uncertainty and risk in the economy. Prices fluctuate because of output volatility and not based on economic fundamentals, causing "signal extraction" problems for investors (Lucas, 1973). This can reduce a country's economic growth rate (Timmer, 2002). Also, output volatility can influence the quantity of investment by increasing savings. Unstable food prices might increase risk throughout the entire economy, fueling more speculative investments and less core investments that could contribute to an increase in physical and human capital over the long run (Timmer, 2002). From a consumer point of view, unstable food prices lead to an increase in precautionary savings in anticipation of further food price increases. These savings linger waiting on food prices to increase. In the absence of sophisticated financial intermediaries, most of these monies are held as cash and are not funneled to the nation's level of core long-term investment (Deaton, 1992).

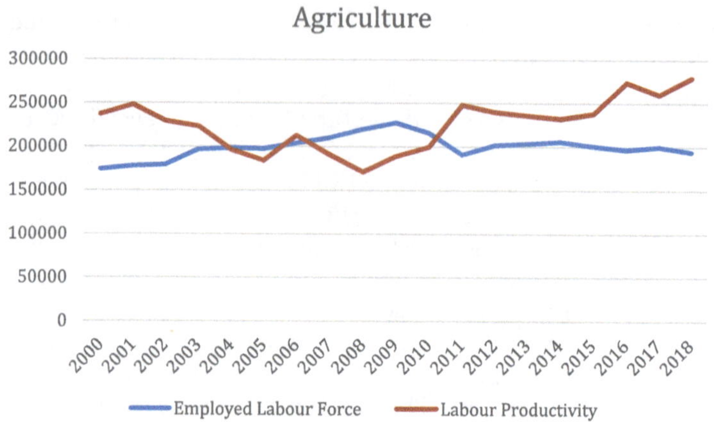

Fig. 3.8 Agriculture productivity and employed labour force (*Source* STATIN)

3.2.1 *The Optimal Level of Employment in the Agricultural Sector*

Figure 3.11 shows agricultural output per person dependent on the number of people employed from 2000 to 2019, while Fig. 3.12 shows total output per number of people employed from 2000 to 2019. Now with more information the same results emerge, and the story remains the same. Three main clusters are identified. The first cluster contains the dot in green to the top of Figs. 3.11 and 3.12 shows that Jamaica produces its maximum agricultural output when it hires between 194,300 and 200,300 people, anything above or below that appears to be suboptimal as it produces less output for the country. Jamaica produces the most output of $279,000 per person when it employs 194,300 people. Jamaica produces its worst output of $171,000 per person when it employed 220,000 back in 2008. Here, more people were hired than optimal and most of this labour gradually left the sector. The cluster in blue to the far right of Figs. 3.11 and 3.12 shows that if Jamaica employs between 204,000 and 227,000 people, the country yields less than $44 billion in total revenues or at most earn $210,000 per person. This is $69,000 less than the optimal per person and less than the $54 billion that the industry can produce in total. In this scenario, Jamaica is hiring 23,000 people that are underutilized in this sector. Jamaica can reduce its productivity loses by hiring labour anywhere between 194,000 and 199,000 or 194,300

Table 3.2 Labour usage monitor: agriculture

Labour usage monitor	*Current*	*Optimal*	*Excess*	*Percent*
Labour	194,300.00	194,300	–	–
Productivity	279,111	279,111	–	–
GDP	54,231,287,236	54,231,287,236	–	–

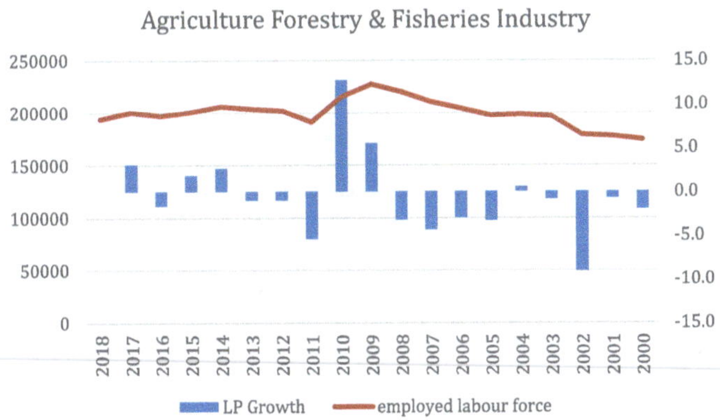

Fig. 3.9 Agriculture productivity growth and employed labour force (*Source* STATIN)

people to be exact, see Table 3.2. The excess people can filter into other employment or training to meet the digital needs of the people (Figs. 3.9, 3.10, 3.11, and 3.12).

3.2.2 Productivity Challenges in Agriculture

Figure 3.13 shows output in tonnes for nontraditional agricultural produce in Jamaica. Total tonnes produced increased from little over 600,000 tonnes to less than 700,000 tonnes in 2019. Other than sugarcane and bananas, Jamaica produces other fruits; 34,000 tonnes of pineapple, 16,000 tonnes of watermelon, 11,000 tonnes of pawpaw and 3,000 tonnes of cantaloupe in 2019. Other than fruits, Jamaica produces mostly vegetables, yams, legumes, condiments, cereals, plantains, potatoes, sorrel and tubers. According to "The estimates of crop production", vegetables is the largest output category. Jamaica produced

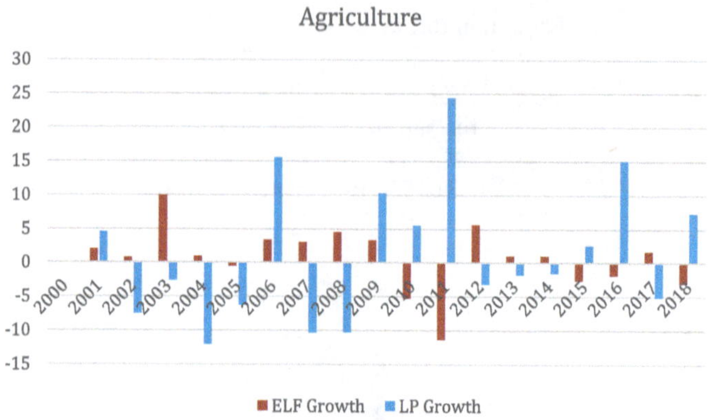

Fig. 3.10 Agriculture productivity and employment growth rates (*Source* STATIN)

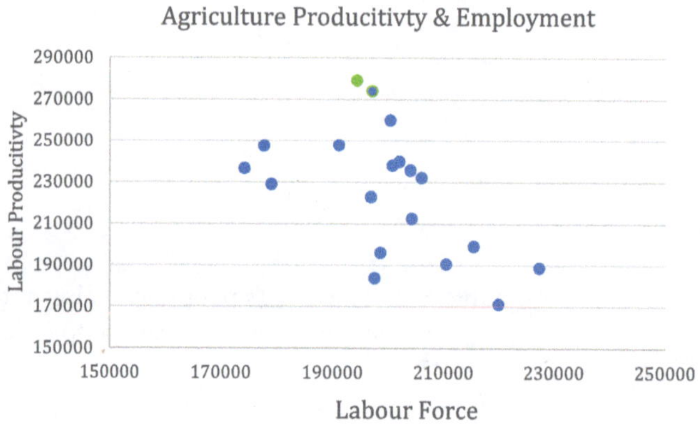

Fig. 3.11 Agriculture productivity and employment scatter (*Source* STATIN)

roughly 242,000 tonnes of vegetables in 2019 coming from just over 16,000 tonnes in 2010 most of which average less than a $1000 USD per tonne. Notice, that production of Jamaica's agricultural crops moves in a similar direction from year to year. Production of everything increases in 2011 coinciding with the increase in productivity displayed in Fig. 3.10. All produce increased in output in 2012 and then declined in 2013 and 2014 increasing again in 2015, declining after but is at the 2015 level in

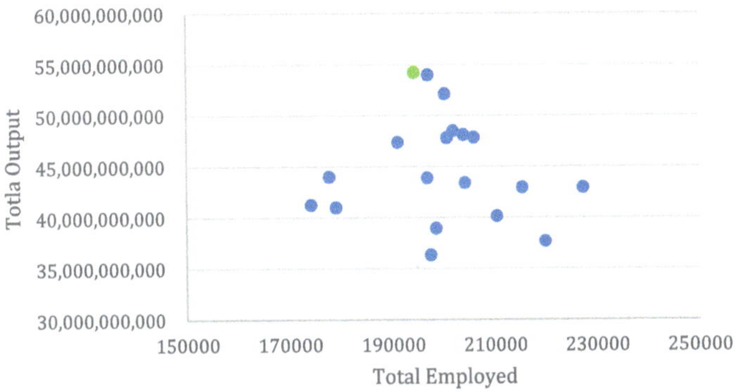

Fig. 3.12 Agriculture GDP and employment (*Source* STATIN)

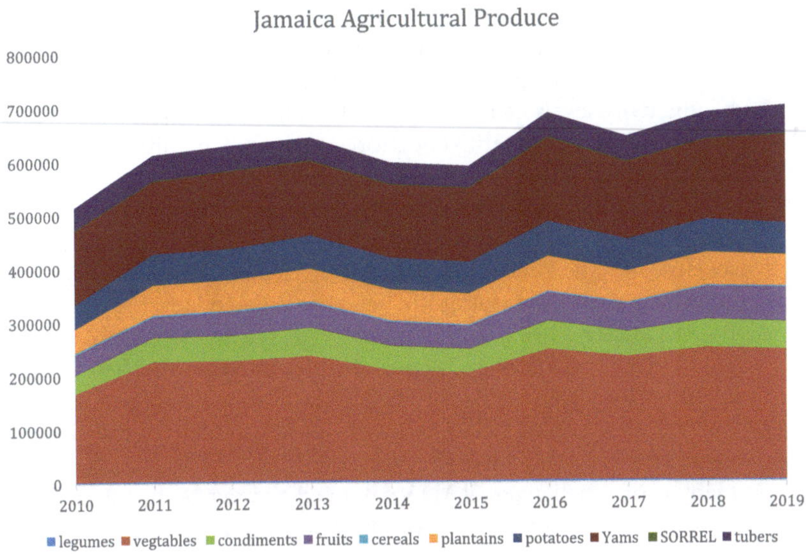

Fig. 3.13 Agricultural produce (2010–2019) (*Source* RADA)

2019, see Fig. 3.13. This highlights that there are structural challenges that affect all the output at the same time. Jamaica's agricultural sector has evolved into a quest to reap the expected economies of scale from large-scale agriculture from a group of small-scale disjointed farms.

Contemporarily, Jamaica's agricultural sector has faced tremendous challenges over the last couple of decades that has resulted in output and productivity volatility. These include but are not limited to (1) Declining competitiveness, (2) Limited staffing and resources for extension services. (3) Gaps in infrastructure, (4) Weaknesses in marketing, (5) Ageing work force, (6) Loss of agricultural lands to urban settlements, (7) Limited capacity for research and development (R&D), (8) Threats to the sustainable use of the island's forestry resources and marine fisheries, (9) Environmental factors and (10) Economic Challenges. These are categorized and expounded upon below.

1. Declining competitiveness of agricultural production has manifested in declining exports and rising imports due to small size of landholdings, high cost of inputs, praedial larceny and limited application of modern technology. Between 2005 and 2018, the agricultural industry in Jamaica received more than $24 million in Foreign Direct Investment (FDI). The year-to-year volatility in productivity levels can be replaced by a targeted approach towards improving the competitiveness of Jamaica's Agricultural Industry incrementally.
2. Limited staffing and resources for extension services. Public sector extension services that normally bridge the gap between research in labs and the practical work of the farmers are sometimes not as effective as they ought to be, especially in developing countries. Timmer (2002) outlines that research is useless unless it is employed on wide scale by farmers. In Jamaica, the Rural Agriculture Development Authority (RADA) serves as the extension, but there are several handicaps with RADA's ability to fully deliver these services. These include lack of budget and resources among many other challenges. RADA will need a larger budget and better expertise if they are to successfully tackle some of the handicaps of the industry.
3. Gaps in key infrastructure include: inadequate maintenance of feeder roads; inadequate irrigation works; lack of sorting, grading, packaging and storage facilities; lack of post harvesting facilities and practices, improper storage, distribution and drainage infrastructure for agricultural areas; poor and inconsistent water supply. People practice different type of water harvesting, storage and distribution but irrigation has been a challenge for Jamaica. The

situation can only be solved long term by the building of dams and proper irrigation infrastructure to service the agriculture parishes like Manchester, Clarendon and St. Elizabeth. There has been debate as to whether the government and private sector can come to a consensus as to who is better able to provide these water and other infrastructure agricultural solutions. According to Timmer (2002), investment in rural irrigation seems to be beyond the private sector due to the innate challenges to coordinate, manage, finance and execute these projects. The rate of return on these investments are in question in many developed countries whose return from agriculture has been declining over the last couple of decades. In Jamaica, unstable agricultural prices, haphazard agricultural industry set up, among many other challenges make it difficult for private firms to consider such investment outlay. The public good issue of excludability due to rivalry arising from supply is also a challenge for private firms. Hence, private investments are normally small and specific instead of large and communal. Small investment like tube well irrigation and water pumped from the river are the main forms of investment, see Timmer (2002). These along with rainwater harvesting are practised in Jamaica. As a result, it is mainly the government who is tasked with the duty to provide sufficient rural infrastructure to facilitate the increase in productivity in the agricultural sector. According to Timmer (2002), if government decides to provide these dams etc., environmental factors for example, the consequences of rerouting water from its natural channel must be taken into consideration. If this infrastructure is not provided by government, then there need to be sufficient incentives for private firms to raise the funds and provide a lucrative return to investors. Water pricing must also be taken into consideration. This will depend on the cost to establish and maintain the infrastructure. According to Timmer (2002), public sector workers are prone to bureaucratic inefficiencies when operating and maintaining the water situation and therefore pricing of the water should depend on the timeliness of the delivery and quality of water services. In reality, most of the water provided for agriculture is either free or subsidized by the government.

4. Weaknesses in marketing include: high levels of informality in marketing and distribution channels, limited market information

services and relatively weak linkages to other economic sectors, industries and non-traditional export markets.

5. Ageing work force: A large percentage of Jamaican farmers are over the age of 40 years old. Young people are not interested in agriculture for two main reasons, the stigma attached to and the rate of return from the industry. As a result, the government as embarked on a path to create a youth in agriculture policy to incentivize youths to join the industry.[5] However, young people are attracted to cannabis farming which they believe have a higher rate of return and a better progressive stigma. Note cannabis sells for anywhere between $50 USD to $3000 USD per pound, when compared to other fruits and vegetables that are less than $5 USD per pound.
6. Loss of agricultural lands to urban settlement and housing development: Some landowners have diversified away from agricultural produce like sugar cane and tobacco into housing and real estate development, thereby reducing the availability of arable lands on the island. Arable land has reduced from 174,000 hectares in 1993 to 120,000 hectares up to 2018. This reflects a 31% reduction in arable lands over the period.[6] The switch from agriculture to real estate was an easy decision for many large-scale landowners. On face value, productivity in the Construction and Installation industry earns ($459,490.42), almost twice as much as agriculture and ($771,095.06) in the Real Estate and Business Services industry, almost three times as much as agriculture.
7. Limited capacity for research and development (R&D). Most of the research in the agricultural sector is normally carried out by universities and then spread to the public. Modern Science and new technologies have revolutionized and enhanced agriculture to generate high income (Timmer, 2002). Investments in these however, are more rewarding in developed countries rather than developing countries. In developing countries, there is little investment in research and development for the agricultural sector. Most innovations are developed by more advanced countries, most research on for example, hybrid seeds, inbred lines,

[5] For further details, see "Youth in Agriculture" on the Ministry of Agriculture and Fisheries website: https://www.moa.gov.jm/content/youth-agriculture.

[6] Authors' calculation from World Development Indicators.

patented chemical formulae are carried out by developed countries (Timmer, 2002). According to Timmer (2002), private firms most times recover a small percentage of the cost of research and as a result, the research must be carried out by the public sector. The ability to scale also negates the motivation for agricultural research from the private sector especially in developing countries. Agricultural innovations and new technology do not get the investment support they need to thrive from the government. Timmer (2002) outlines that, only in countries that can facilitate the investment in agricultural research can agriculture thrive.

8. Threats to the sustainable use of the island's forestry resources, including natural and man-made hazards, soil erosion and illegal logging. Threats to the long-term development of Jamaica's marine fisheries including over-harvesting; habitat destruction and pollution; and incursions by fishing vessels from other countries including Colombia and Honduras.
9. Environmental factors like global warming, natural disasters like hurricanes, storms, drought, floods and other might have an impact on the productivity of the agricultural sector output. In most agricultural-oriented areas across the island, there is water issues. Drought sometimes, flood sometimes without a happy medium adverse weather conditions floods, droughts. Other environmental issues including soil erosion from over-cultivation and inadequate soil conservation techniques in hillside farming; use of chemical fertilizers; impact of natural hazards; and deforestation resulting from clearing of hillsides, illegal settlements, monoculture farming and uncontrolled harvesting of trees all have impact as well.
10. Economic Challenges and Major Impediments include shocks to world prices for key inputs such as fertilizers, feed and fuel, all of which are mainly imported. Changes in export market prices and trading regimes, praedial larceny, use of inappropriate technologies, difficulties in titling and transfer of lands, high cost of and limited access to capital. Timmer (2002) also noted that a lack of economic freedom, poor institutions, ineffective economic policies and political instability can also reduce the productivity of the agricultural sector.

3.2.3 Overcoming Challenges to Agricultural Industry

3.2.3.1 Agricultural Policy

Despite the above challenges, there is room for Jamaica to improve the output of its agricultural sector significantly. Furthermore, evidence from Timmer (2002) among others suggest that productivity in the agricultural sector is important to facilitate sustainable increase in productivity in the industrial sector. To do so, Timmer (2002) outlines that coordinated policy in four areas are needed: The macroeconomic environment, Technology, Physical capital deepening, human capital deepening, and incentives for farmers to produce.

1. Macro environmental factors

Macro environmental factors include balance of payments internal and external accounts, macroeconomic price stability, a competitive exchange rate in favour of exports and a transparent and unbiased trading regime. Timmer (2002) highlighted that agricultural ministers rarely address these issues, which appear to be true in the case of Jamaica and many other SIDs. Schuh (1976), Timmer et al. (1983) suggest that the macro environment must be supportive of progression in the agricultural sector; if not, it will remain stagnated with limited growth such has been the case in Jamaica.

2. Technology

The right technology is needed to increase the crop yield per acre or unit of land. Johnson (1997) reasons that agricultural output has transformed from subsistence level of using traditional techniques to integrating technology and modern techniques that can enable the production of a vast amount of output for commercial sale. The rate of technological advancement of the agricultural sector is crucial to facilitate this growth. From a theoretical point of view, an increase in the earnings from agriculture should increase the imports of new technology and physical capital. In more developed countries, agricultural export is characterized by high value added compared to manufacturing export where there is a huge import content in the latter. In Jamaica, this is not so.

3. Physical capital deepening

Savings from the agricultural sector make more investing cheaper and facilitate capital deepening in the industrial sector thus facilitating more growth. Jamaica has a low savings rate and therefore has not been able to accumulate capital at a pace that would facilitate the rapid expansion of agriculture. Savings is less productive if accumulated by the government rather than private entities. Public funds when used for infrastructural expansion through the provision of public goods can improve the productiveness and the profitability of the private sector. Timmer (2002) outlines that the agricultural sector requires a market-oriented strategy capable of stimulating rapid technological change. Any attempt to understand the impact of agriculture on economic growth should be done in the context that identifies a positive relationship between agriculture and economic growth (Johnson, 1997).

4. Human Capital deepening

Growth in the productivity of the agricultural sector can influence rural income and therefore rural educational levels, see Chai (1995) and Birdsall et al. (1995). The process is a revolving, mutually beneficial, symbiotic relationship as the increase in education and income should in turn improve agricultural productivity (Jamison & Lau, 1982). If agricultural productivity is high, then migration of children of farmers is less painful and more rewarding.

5. Incentives

People are driven by money and the best way to get them to produce is if they are getting a good return. Agriculture in Jamaica is no different. Fundamentally, for Jamaica's agriculture sector to progress, it must correct all its handicaps and look to value creation through the Beanstalk theory, High growth plants, nutraceuticals and medicinal herbs ultimately creating an avenue for patented products.

3.3 Agriculture's Correlation with Other Industries in Jamaica

The correlation between the labour productivity of the Agriculture, Forestry and Fisheries Industry and the other labelled industries in Jamaica are illustrated in Table 3.3. The correlation analysis shows that Agriculture Forestry and Fishing has no significant correlation with any other industry in Jamaica. Timmer (2002) suggests that balance growth is needed between the agricultural sector and the industrial sector for a country to develop. Productivity in the agricultural sector is also important to influence rural education levels, which then increases overall productivity further, see Timmer (2002), Chai (1995), Birdsall et al. (1995), and Jamison and Lau (1982). Other industries within the goods producing sector like Construction is positively correlated with Mining and Quarrying and Wholesale and Retail Trade where Wholesale and Retail Trade is correlated with Mining and Quarrying and Manufacturing. From a service point of view, the Financial and Insurance industry is correlated with Electricity, Gas and Water as well as Transport and Storage. Real Estate is correlated with Wholesale and Retail. The services industries are discussed in the next chapter. Despite the lack of correlation, development of the agricultural sector is very important for a country to achieve holistic economic growth (Johnson, 1997). Timmer (2002) outlines that no country has ever achieved long-run rapid economic growth without first achieving food security.

About 42% of the money that tourists who visit Jamaica spends is on food. The tourism industry demands approximately 350 billion worth of supplies from the agricultural industry but local farmers are only able to supply a fifth or 20%.

3.4 The Future of Manufacturing on the Island

Over the last 20 years, production in Jamaica's manufacturing sector has fallen by more than 5.5 billion or 10%, been negatively impacted by many factors including cost of inputs such as gas and oil prices. According to the National Five-year manufacturing growth strategy, Jamaica's manufacturing sector declined from a share of more than 20% of GDP in 1985 to less than 10% of GDP presently. Total manufacturing exports from Jamaica was valued at US$624,565 or 4.2% of Jamaica's GDP in 2018. Traditionally, the Jamaican manufacturing industry was centred around

Table 3.3 Industry correlation matrix

	A,F&F	*M&Q*	*Manu*	*E&W*	*Cons*	*W&R*	*H&R*	*T&C*	*F&I*	*RE&BA*	*P&G*
Mining & Quarrying	−0.1	1.0									
Manufacture	−0.1	0.4	1.0								
Electricity & Water Supply	0.2	−0.1	−0.2	1.0							
Construction	−0.4	0.6	0.3	0.1	1.0						
Wholesale & Retail Trade; Repairs; Machinery	−0.2	0.5	0.5	0.4	0.5	1.0					
Accomodation and Food Service Activities	0.2	0.0	−0.1	0.2	0.0	0.4	1.0				
Transport & Storage	−0.1	0.3	0.0	0.5	0.4	0.5	0.2	1.0			
Financial & Insurance Activities	0.0	−0.1	0.0	0.6	0.0	0.3	−0.1	0.6	1.0		
Real Estate R & BA	−0.2	0.3	0.4	0.2	0.5	0.7	0.2	0.3	0.4	1.0	
Government Services	0.0	0.1	−0.1	0.2	0.1	0.0	−0.3	0.3	0.4	0.3	1.0
Other Services	0.1	0.2	0.1	0.6	0.3	0.7	0.6	0.5	0.4	0.7	0.1

the processing of sugarcane, rum and similar products which account for about 20% of total manufacturing exports in 2018. Gradually over time Jamaica diversified in the manufacturing of other non-traditional goods which account for 80%, the higher portion of the total manufacturing exported goods on the island. Mineral fuels of US$279 million is Jamaica's most valuable manufactured commodity exported. Followed by processed food valued at US$139.5 million in 2018, Jamaica exports US$66 million worth of manufactured beverages excluding rum, US$37 million worth of chemicals including ethanol, US$24 million worth of crude materials, animal and vegetable oil, manufactured goods, machinery and transport equipment add up to less than US$10 million in total (Fig. 3.14).

Along with declining contribution to GDP, between 2000 to 2018, Jamaica's manufacturing sector demonstrates a relatively constant value of output per person of approximately $649,707.95, see Fig. 3.15. Research conducted by the World Bank indicates that productivity growth in manufacturing was driven by efficiency gains in countries with higher capacity for innovation. Alam (2008) indicated that countries in which the manufacturing industries employed a large proportion of high skilled-workers, produced ICT goods or used a relatively large amount of ICT capital had higher productivity growth rates relative to low technology groups.

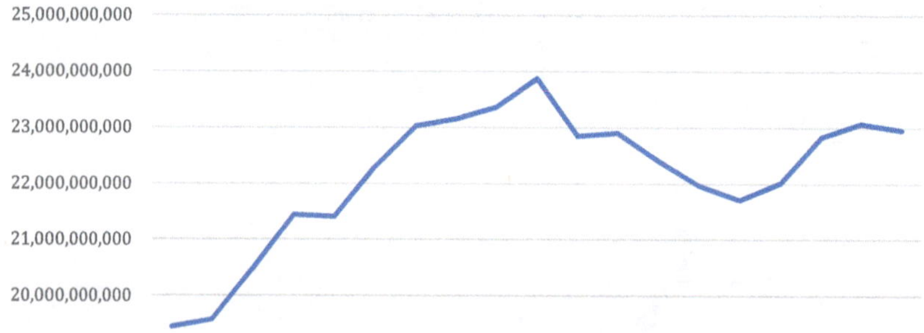

Fig. 3.14 Manufacturing GDP

If Jamaica is to transform its economy, increased human capital development will play a major role in the manufacturing industry (Figs. 3.16 and 3.17).

Jamaica's National Export Strategy (NES) outlines Jamaica's plans to focus on Agro-processing and light manufacturing to boost manufacturing productive output in the industry. Light manufactured tools include electric, metals, paper, chemical products and building tools, while agro-processing includes manufacturing of food-based products. According to the report, the NES aims to achieve a 25% growth in total exports and US$2.5 million export in light manufacturing goods by 2019. Jamaica exported roughly $1.5 million worth of light manufactured goods in 2018. The strategy outlines that Jamaica will need to innovate to shift away from basic manufacturing to more technical technology-driven value-added production. The target was not met for 2019.

The NES identifies 6 main product segments of focus based on Jamaica's comparative advantage and niche areas. These include Agro-processing, electronics and medical devices have been earmarked as key areas of focus light manufacturing. Cosmetics, furniture, pharmaceuticals, packaging.

Global manufacturing efficiency is driven by 12 pillars broken down into two groups, market forces and government forces. Market forces

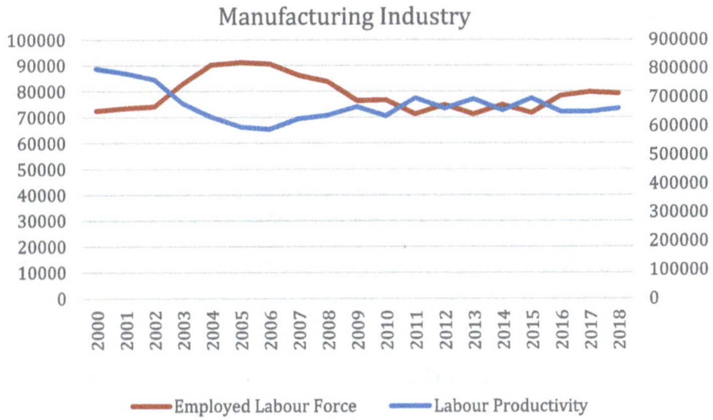

Fig. 3.15 Manufacturing employed labour force and productivity (*Source* STATIN)

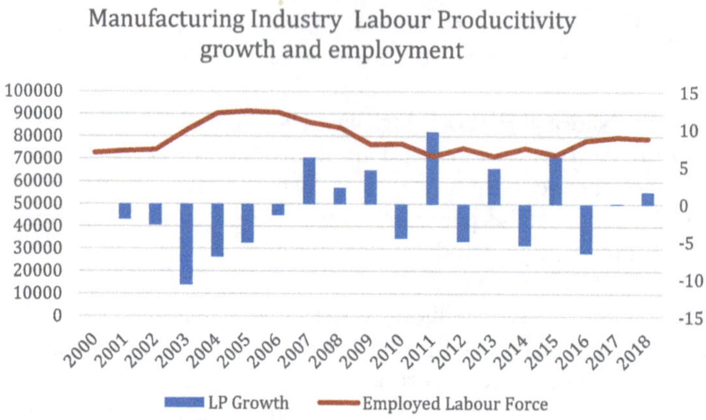

Fig. 3.16 Manufacturing productivity and employed labour force growth (*Source* STATIN)

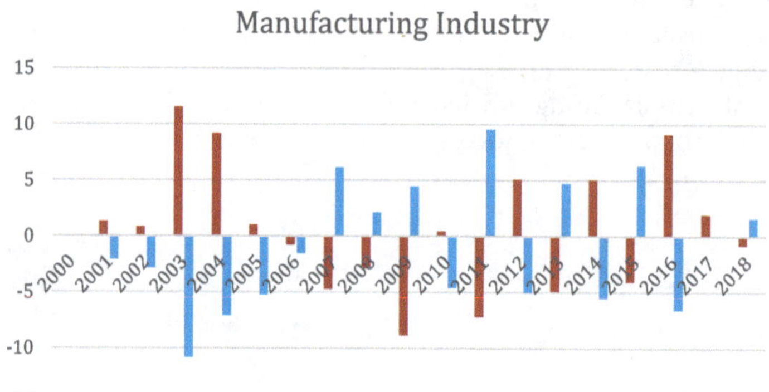

Fig. 3.17 Manufacturing employed labour force and productivity growth rates (*Source* STATIN)

include talent, cost competitiveness, workforce productivity, supply network, local markets and attractiveness, while government forces include legal and regulatory system, physical infrastructure, economic trade, financial and tax system, innovation policy and infrastructure, energy and the healthcare system, all of which currently face challenges in Jamaica.

A SWOT analysis outlined a list of Jamaica's strengths, weaknesses, opportunities and threats. Strengths include good road infrastructure for transportation of goods, competitive labour cost in accessing unskilled and semi-skilled labour, access to raw materials for production and existing intellectual property legislative framework.

The manufacturing sector faces severe challenges in unearthing its productive potential including the cost of doing business which includes the cost of bureaucracy, legal and legislative framework. Crime and corruption are also noticeable hindrances to doing effective business across the island. Jamaica as well has depleting physical infrastructure; the manufacturing infrastructure is outdated, and successive budgets have focused on debt repayment with little remaining for capital and infrastructural expansion. Jamaican markets have poor supply networks, high electricity per kilowatt hour and human capital requirement. Reducing the cost of energy is necessary to facilitate intense manufacturing to improve output per person. Jamaica must explore cheaper ways of producing electricity either by water, solar or wind. Firms can partner with the universities to produce students with the necessary skills to manufacture goods. Some of the manufacturing processes are import-dependent. This coupled with poor and insufficient industry and production standards make manufacturing a challenge in the small island.

Despite the weaknesses, there are tremendous opportunities through the shift in global manufacturing patterns and the development of new technologies that might allow Jamaica to pursue its light manufacturing goals. These however continue to be threatened by limited domestic economies of scale and lack of critical Research and Development.

3.4.1 The Optimal Level of Employment in the Manufacturing Sector

Jamaica's manufacturing employed labour force ranges from a low of 71,000 people to a high of 91,000 people. Producing output per worker of $588,000 to a high of $988,000, see Fig. 3.18, a scatter plot that shows the relationship between the number of person's employed and total output, while Fig. 3.19 shows the relationship between total employed and labour productivity. The yellow dot on both curves shows the level of employment that produces the highest level of output,

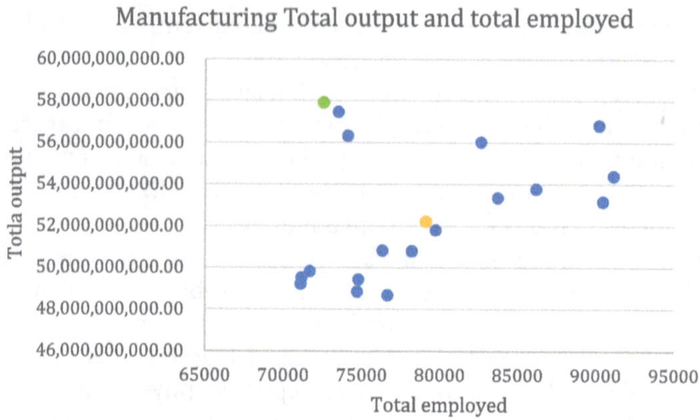

Fig. 3.18 Manufacturing total output and employment (*Source* STATIN)

while the green dot on both curves shows the level of employment that produces the highest level of productivity. Observe the yellow dot on Fig. 3.18, which shows the highest level of output, while the green dot shows low output, while on Fig. 3.19 the green dot shows high labour productivity while the red dot shows low labour productivity. As more people are employed, each unit of labour becomes less and less productive. In fact, the maximum level of productivity is achieved when the least amount of people is employed, the green dot on Fig. 3.19 indicating that much of the employed are redundant. Current level of employment of 79,100 people each with an average productivity level of $660,311 JMD. Optimal productivity achieved is $798,000. Current value-added GDP is roughly $58 billion, which would increase by 21% to $63 billion if everyone were as productive as they can be, see Table 3.4. At this rate of productivity, it would require only 65,500 to produce the current level of GDP. The agricultural industry is currently hiring 17% more people than what is required if the country was producing at optimal level of labour productivity. As a result, the remaining 13,700 people currently working in the manufacturing sector are under-employed and under-productive.

3.5 Mining and Quarrying

Mining and Quarrying is the largest goods producing sector in Jamaica. At $2,293,343.91 per person employed, it is also employing the most

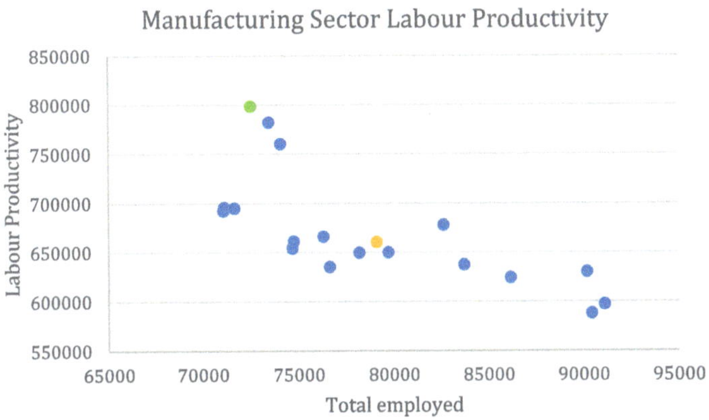

Fig. 3.19 Manufacturing labour productivity and employment (*Source* STATIN)

Table 3.4 Labour usage monitor: manufacturing

Manufacturing				
Labour usage monitor	*Current*	*Optimal*	*Excess*	*Percent (%)*
Labour	79,100	65,417	13,683.18	17
Productivity	660,311	798,428	−138,116.68	−21
GDP	52,230,612,258	63,155,641,430	−10,925,029,171.77	−21

Optimal Labour = Current GDP/Optimal Productivity 72,500 is the optimal, now the country is employing 79,100, 6600 more people than what is necessary to produce the optimal

productive labour among the goods producing industries. The main metallic mineral resources mined in Jamaica are bauxite, cobalt, copper, gold, iron, lead, manganese, silver, titanium and zinc, producing Limestone, Sand and Gravel Shale, Pozzolan, Gypsum, Silica, Clay, Marble, Marl and Fill. Jamaica's most abundant mineral is Limestone, 98% CaCO3 high purity grade. Since the onset of the Global Financial Crisis, GDP in the sector has not recovered to post-crisis levels, see Fig. 3.20. Up to 2019, Mining and Quarrying continues to be the most productive industry in the Goods Producing Sector in the Jamaican economy in terms of the value of output per employed worker. Output per worker was once as high as $5,000,000 JMD per person back in 2005, since then, output per person demonstrates some level of seasonal decline. See Fig. 3.21. Between 2005 and 2018, roughly $838.3 Million US or 24% of the Foreign Direct Investment pumped into Jamaica for the purposes of Mining and Quarrying. Over that time, Manufacturing, Agriculture and its sub- groups, Forestry and Fishing and Construction industries have not been as productive as the Mining Industry. Mining and Quarrying took a huge blow when one of the islands' largest bauxite companies ceased operation in 2019. Value-added output in the sector fell by more than 20% in 2020. The industry has even more potential if Jamaica were to engage in vertical diversification into higher value-added production from the output of this industry. These processes are energy-intensive, but again, the high cost of energy minimizes the island's ability to engage in manufacturing.

On average, the conversion rate of two of the three major inputs used in refineries to produce a tonne of alumina increased in 2019. Lower efficiency was registered for bauxite to alumina, and caustic soda to alumina. The Bauxite/Alumina conversion rate was 2.73 tonnes of bauxite per tonne of alumina relative to 2.69 tonnes in 2018, while the Caustic Soda/Alumina conversion rate was 0.160 tonnes of caustic soda per tonne of alumina relative to 0.141 tonnes in 2018. Less fuel was used to produce one tonne of alumina in 2019 relative to 2018, signaling an improvement in the conversion rate of fuel to alumina, the third major input (Figs. 3.22 and 3.23).

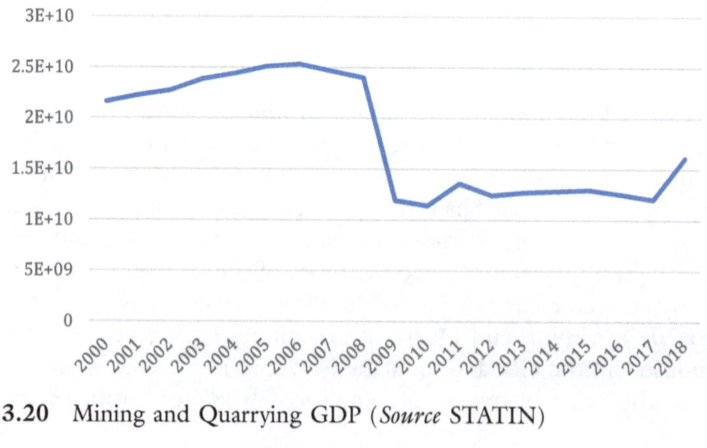

Fig. 3.20 Mining and Quarrying GDP (*Source* STATIN)

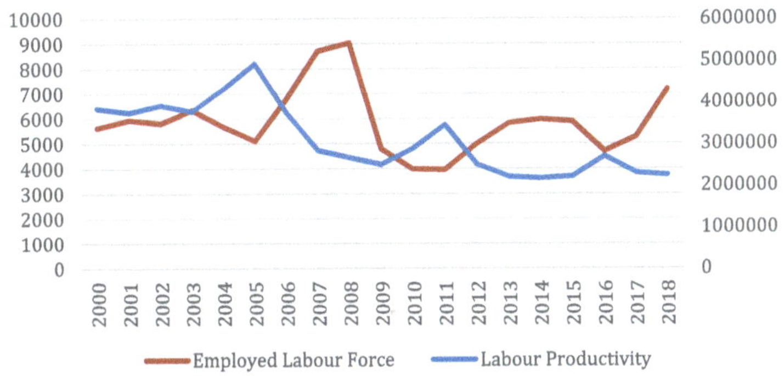

Fig. 3.21 Mining and Quarrying labour productivity and employment (*Source* STATIN)

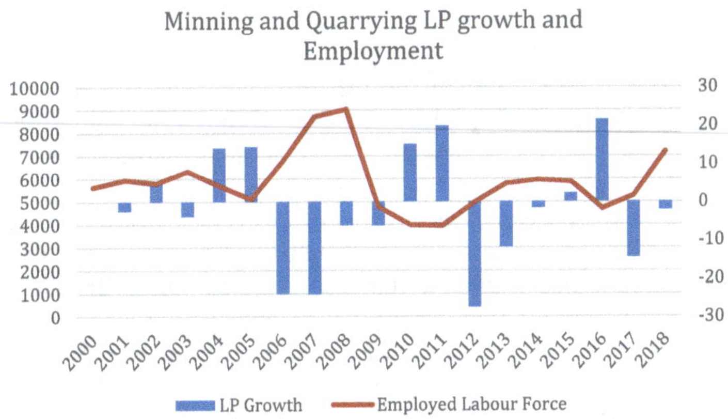

Fig. 3.22 Mining and Quarrying labour productivity growth and employment (*Source* STATIN)

3.5.1 *The Optimal Level of Employment in the Mining and Quarrying Sector*

Employment in the Mining and Quarrying sector has grown from 4000 people 20 years ago, to over 7,000 people today. Producing output per worker as low as $2.2 million to 588,000 to a high of $4.9 million. As usual, the scatter plots show the relationship between the number of persons employed and total output, and the relationship between total

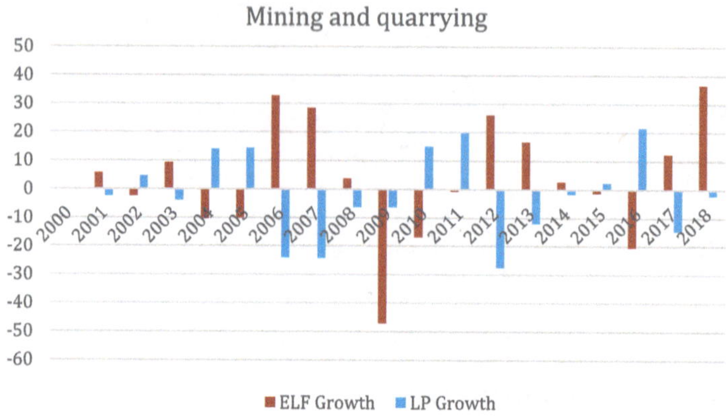

Fig. 3.23 Mining and Quarrying employment and productivity growth rates (*Source* STATIN)

employed and labour productivity, see Fig. 3.21. The red dot on both curves shows the level of employment that produces the highest level of output, the orange dot shows the current level of output and productivity, while the green dot on both curves shows the level of employment that produces the highest level of productivity. Current level of employment is 7,175 people each with an average productivity of level of $2.225 million JMD, see Fig. 3.24. The highest level of output $25 billion can be produced by just 3,271 people if they produce at optimal productivity level of $0.9 million, see Fig. 3.25. At optimality, labour in this sector could fall by as much a 54% and still produce the optimal. If the current labour is employed at its highest level of productivity, the economy could produce up to $35 billion, 19 billion or 119% more than it is currently producing now, see Table 3.5 on Labour usage monitor: Mining and Quarrying.

3.6 Construction to Industrialize

Jamaica is a developing country which means that construction is way below its production possibility frontier. The construction industry is crucial to create structures to sustain economic activity. Roads, railways, waterways, drainage systems, commercial and residential buildings are forms of constructions that require constant upkeep to enable a country

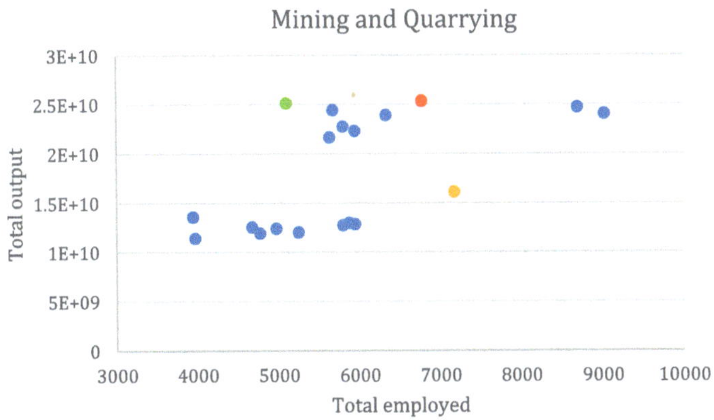

Fig. 3.24 Mining and Quarrying total output and employment (*Source* STATIN)

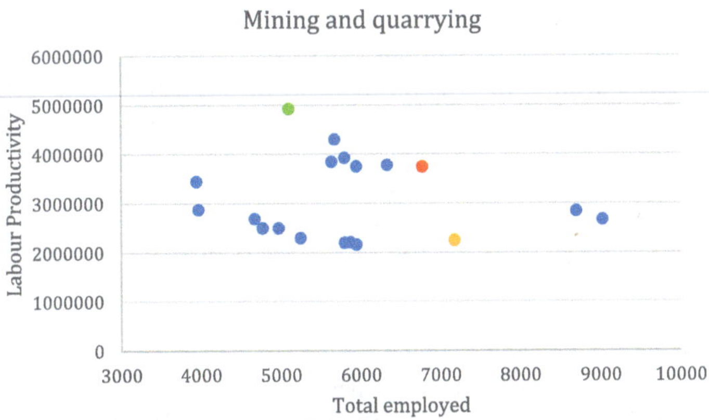

Fig. 3.25 Mining and Quarrying productivity and employment (*Source* STATIN)

Table 3.5 Labour usage monitor: Mining and Quarrying

Labour usage monitor	*Current*	*Optimal*	*Excess*	*Percent (%)*
Labour	7,175	3,271	3,903.56	54
Productivity	2,244,554	4,922,804	−2,678,249.92	−119
GDP	16,104,674,946	35,321,118,137	−19,216,443,191.33	−119

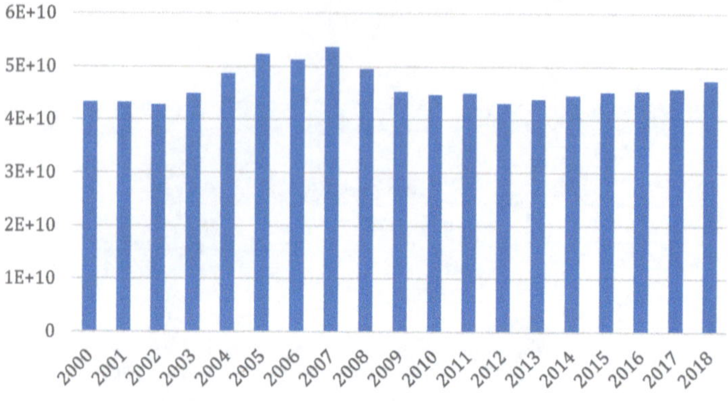

Fig. 3.26 Construction GDP (*Source* STATIN)

to transition smoothly on its developmental path. Jamaica's construction industry is the 4th least productive industry of all the industries, and second least productive goods industry outside of agriculture. Total output ranges from $433 billion in 2000 to $473 billion in 2018, see Fig. 3.26. Jamaica's construction industry has been gradually increasing over the last two years but is less than what it was prior to the Global Financial Crisis in 2008. Albeit the industry has faced some challenges over the years including human resource capabilities and competitiveness which have restrained the rapid advancement of the industry. Measures to incorporate technology, and those that promote sustainable development are needed at the forefront.

Plans to improve the construction industry in Jamaica are laid out in Jamaica's Construction Industry Policy Paper 2014.[7] It outlines that the business environment must be made more enabling by the public sector; better procurement procedures and legislation, more opportunities to expand, more integration between industries, better occupational safety and health, better and more effective monitoring, appraisal and evaluation systems.

Employment and labour productivity appear to be negatively related as employment increase labour productivity falls and vice versa, see Fig. 3.27. Employed labour force, growth and productivity growth

[7] https://japarliament.gov.jm/attachments/article/341/CONSTRUCTION%20INDUSTRY%20POLICY.pdf.

display similar oppositive movement, see Figs. 3.28 and 3.29. The government has outlined that they will attempt to solve the human deficiency problem first through training of workers, to improve the capacity of the public sector, improving their project management skills, stable industrial relations, gender equality, facilitate those with special needs. Sustainable construction is necessary to cope with climate change planning and to facilitate the infusion of renewable energy.

Fig. 3.27 Construction employment and productivity (*Source* STATIN)

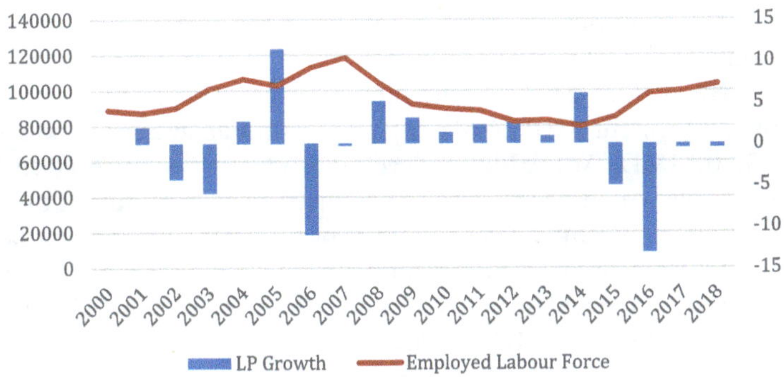

Fig. 3.28 Construction productivity growth and employment (*Source* STATIN)

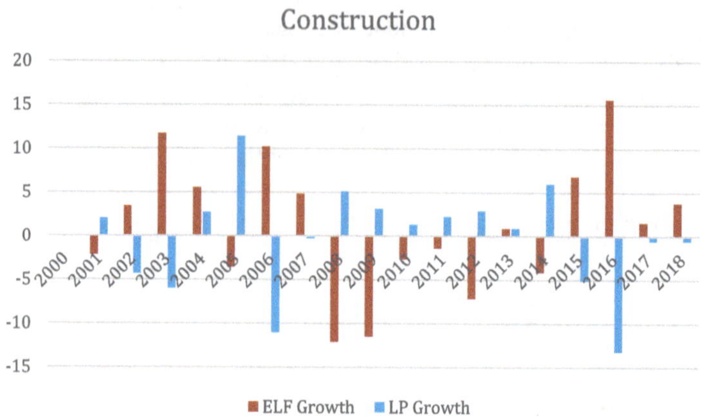

Fig. 3.29 Construction employment and productivity growth rates (*Source* STATIN)

3.6.1 *The Optimal Level of Employment in the Construction Sector*

Employment in the construction sector has fluctuated between a low of 79,600 people, to a high of 118,450 people, see Figs. 3.30 and 3.31. Producing output per worker as low as $452,000 to a high of $550,000. As usual, the scatter plots show the relationship between the number of persons employed and total output, and labour productivity. The red dot on both curves shows the level of employment that produces the highest level of output, the orange dot shows the current level of output and productivity, while the green dot on both curves shows the level of employment that produces the highest level of productivity. Current level of employment is 103,600 people each with an average productivity of $463,000 JMD. At the highest level of productivity of $560,000, only 84,550 people would be required to produce the current output of $47.3 billion. If labour is employed at optimal productivity, the current labour force could produce up to $58 million, about 10.6 billion or 23% more. At optimality, labour in this sector could fall by as much as 18% and still produce the optimal, see Table 3.6.

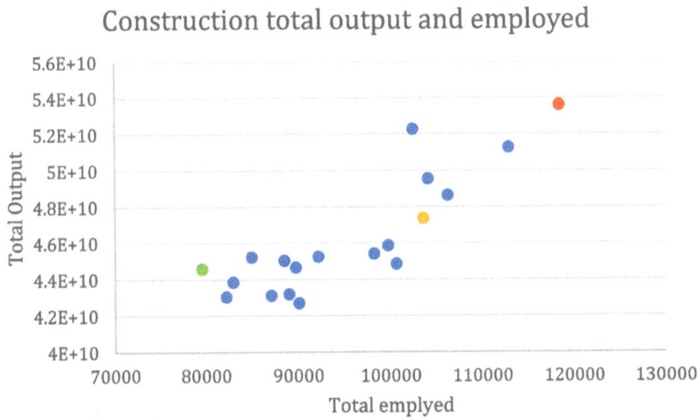

Fig. 3.30 Construction total output and employment (*Source* STATIN)

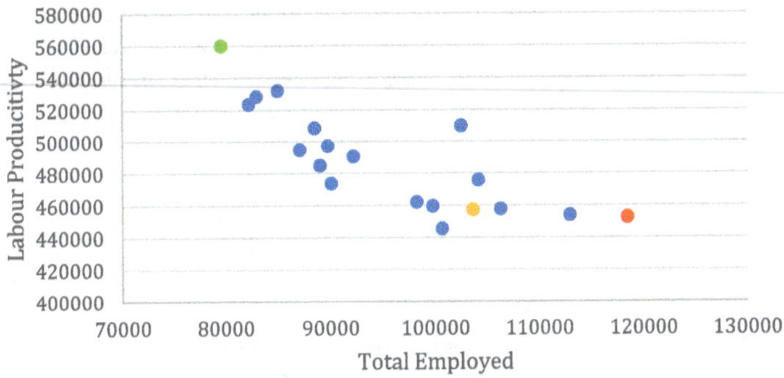

Fig. 3.31 Construction labour productivity and employment (*Source* STATIN)

Table 3.6 Labour usage monitor: construction

Construction				
Labour usage monitor	*Current*	*Optimal*	*Excess*	*Percent (%)*
Labour	103,600	84,556.69	19,043.31	18
Productivity	457,154	560,112	−102,957.33	−23
GDP	47,361,172,954	58,027,552,484.40	−10,666,379,530.26	−23

References

Alam, S. S., Khatibi, A., Ismail Sayyed Ahmad, M., & Bin Ismail, H. (2008). Factors affecting e-commerce adoption in the electronic manufacturing companies in Malaysia. *International Journal of Commerce and Management, 17*(1/2), 125–139.

Birdsall, N., Ross, D., & Sabot, R. (1995). Inequality and growth reconsidered: Lessons from East Asia. *The World Bank Economic Review, 9*(3), 477–508.

Carter. (1997). *Why workers won't work: The worker in a developing economy: A Case Study of Jamaica*. https://books.google.com.jm/books?id=6hNGAAAAYAAJ&q=why+workers+wont+work&dq=why+workers+wont+work&hl=en&sa=X&redir_esc=y

Chai, C. P. (1995). *Rural human capital an economic growth in developing countries*. Doctoral dissertation, Harvard University.

Clarke, S. (1991). New deal regulation and the revolution in American farm productivity. A case study of the diffusion of the tractor in the Corn Ielt, 1920–1940. *The Journal of Economic History, 51*(1), 101–123.

Deaton, A. (1992). *Understanding consumption*. Oxford University Press.

Jamison, D. T., & Lau, L. J. (1982). *Farmer education and farm efficiency*. Johns Hopkins University Press.

Johnson, D. G. (1997). Agriculture and the Wealth of Nations. *The American Economic Review, 87*(2), 1–12.

Johnston, B. F., & Mellor, J. W. (1961). The role of agriculture in economic development. *The American Economic Review, 51*(4), 566–593.

Kuznets, S. (1954). Introduction. In *Trends and cycles in capital formation by United States Railroads, 1970–1950* (pp. 1–9). NBER.

Landmann, O. (2005). *Employment, productivity and output growth*. International Labour Office.

Lewis, W. A. (1954). Economic development with unlimited supplies of labour. *The Manchester School, 22*, 139–191.

Lucas, R. E. (1973). Some international evidence on output-inflation tradeoffs. *The American Economic Review, 63*(3), 326–334.
Mankiw, N. G. (2010). *Intermediate Macroeconomics: Abridged*. Macmillan.
Nordhaus, W. D. (2005). *The sources of the productivity rebound and the manufacturing employment puzzle* (NBER Working Paper No. 11354).
Romer, C. D. (1999). Changes in business cycles: Evidence and explanations. *Journal of Economic Perspectives, 13*(2), 23–44.
Schuh, G. E. (1976). The new macroeconomics of agriculture. *American Journal of Agricultural Economics, 58*(5), 802–811.
Syrquin, M. (1988). Patterns of structural change. *Handbook of Development Economics, 1*, 203–273.
Timmer, C. P. (2002). Agriculture and economic development. *Handbook of Agricultural Economics, 2*, 1487–1546.
Timmer, C. P., Falcon, W. P., & Pearson, S. R. (1983). *Food policy analysis* (Vol. 1983, pp. 1–301). Johns Hopkins University Press.
UN World Development Survey.
Vo, B., Kongar, E., & Suárez Barraza, M. F. (2019). Kaizen event approach: a case study in the packaging industry. *International Journal of Productivity and Performance Management, 68*(7), 1343–1372.
Yildirim, Z. (2015). Relationships among labour productivity, real wages and inflation in Turkey. *Economic Research-Ekonomska Istraživanja*.

CHAPTER 4

Productivity and Labour Usage in the Service Providing Sector in Jamaica

Chapter 3 illustrated that the Goods Producing Industries (GPI) in Jamaica are employing more labour than what is necessary and are producing less output than they could if labour was more efficiently employed. Chapter 4's aim is to ascertain if the same exists for the Service Providing Industries (SPI) in the country. Rutkauskas and Paulavičienė (2005) outline that the labour component of the service sector is more important than the goods producing given that services are more interpersonal. Estimating the efficiency of labour employed is important for Jamaica because services account for approximately 70% of total economic activity, roughly US$10.5 billion in annual income, more than twice as much as the Goods Producing Industries (GPI). Collectively, the SPIs employ roughly three quarters of Jamaica's employed labour force (800,000 people). This is likely to increase given that a lot of emphasis is being placed on the expansion of service provision in the country. For example, there are plans to expand the Business Process Outsourcing industry which seeks to add another 20–30 thousands jobs to the economy by 2025.[1] There are also plans to expand the Accommodation and Food services by another 12,000 rooms by 2024, which will

[1] See link for more details https://www.jamaicaobserver.com/business/us1-billion-targeted-for-bpo/.

A. Haughton and W. Ivey, *Overcoming Productivity Challenges in Small Countries*, https://doi.org/10.1007/978-3-031-23301-2_4

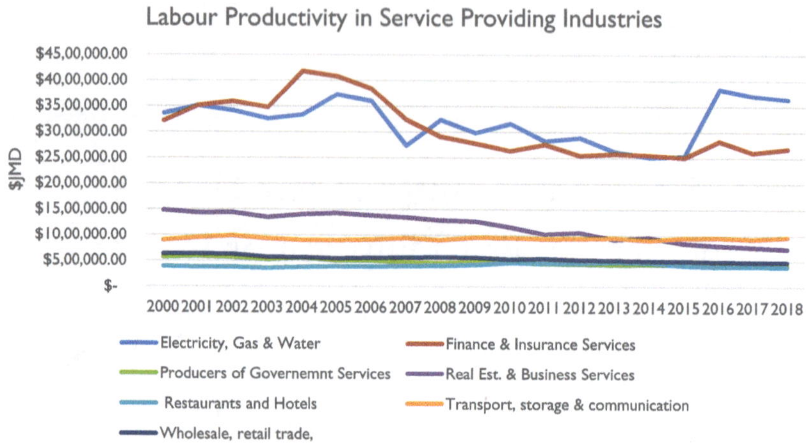

Fig. 4.1 Productivity in the service providing sector (*Source* STATIN)

equate to more jobs as well.[2] Mark (1982) noted that the service sector has the most potential but is characterized by many challenges.

Output per worker in the Service Providing Industries is approximately 2.5 times greater than the Goods Producing Industries; JMD $ is 9.3 million compared to about $3.6 million. The service industries account for approximately 72% of the total share of the value of output per worker compared to 28% in Goods Producing industries.[3] Figure 4.1 shows output per person in each industry of the service providing sector.[4] The Financial and Insurance industry as well as Electricity, Gas and Water have the highest level of labour productivity over the last 20 years. The others have remained relatively low or decreasing.

Producers of Government Services, one of the largest employers on the island, has a low output per person valued at approximately JMD $413,000 (less than $3000 USD per annum). Outside of Producers of Government Services, the Wholesale and Retail industry hire most Jamaicans. Their labour is predominantly unskilled (with output per

[2] Jamaica Information System (2019) 12,000 New Hotel Rooms Over Five Years https://jis.gov.jm/12000-new-hotel-rooms-over-five-years/.

[3] Using 2018 figures, which were the latest available.

[4] Based on Jamaica Productivity Centre's (JPC), calculations using data are provided by the Statistical Institute of Jamaica (STATIN).

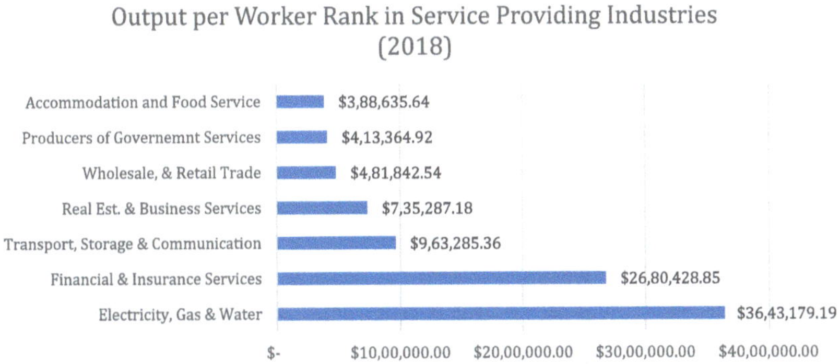

Fig. 4.2 Output per worker rank in service providing industries (*Source* STATIN)

worker of JMD $481,000, little over $3000 USD per annum) see Fig. 4.2. Table 4.1 shows summary statistics for the industry-level productivity; the highest productivity growth on average are labelled in green and those that are lowest in red. Real Estate grows at a rate 4.9% the highest among them, followed by Electricity Gas and Water 1.6%, Transportation 0.3%, Accommodation and Food Services grows at a rate of 0.2%, and Financial and Insurance Services −0.9% per annum over the past 20 years. Producers of Government Services is −1.51%, while Wholesale and Retail Trade is −1.53% which is the least productive.

It is essential to understand labour productivity and the efficient use of labour in each industry to know how to move each up the productivity ladder. This will ultimately benefit the overall economy and provide lessons to other countries and regions, especially when global and local productivity growth rates have been slowing down. This chapter analyses the optimal labour usage and the excess labour employed, the optimal GDP and the excess GDP that could be produced if labour was used effectively as well as the optimal and the excess labour productivity that could be achieved from the existing labour if they were efficient in the service providing industries, see Eq. 11 to Eq. 16 in Chapter 0. The rest of the chapter is outlined as follows: Sect. 4.1 looks at Accommodation and Food Services, Sect. 4.2 looks at Wholesale and Retail Trade, Sect. 4.3 analyses the Electricity Gas and Water Supply, Sect. 4.4 dives into

Table 4.1 Summary statistics: productivity in the service providing sector

	Mean	*Median*	*Min*	*Max*	*Sum*	*Stand. dev*	*Sample variance*	*Range*	*Standard error*
Electricity gas and water supply	1.61	−2.69	−24.06	51.15	27.37	15.86	251.50	75.22	3.85
Accommodation and food	0.16	0.35	−8.55	8.83	2.78	4.42	19.53	17.38	1.07
Transport and storage	0.26	0.77	−6.00	5.50	4.36	3.62	13.08	11.50	0.88
Wholesale and retail trade	−1.53	−1.35	−8.81	2.42	−26.06	2.62	6.88	11.23	0.64
Financial and insurance activities	−0.88	−2.30	−15.48	20.21	−14.99	8.78	77.11	35.69	2.13
Real estate and other business	4.90	5.26	−3.27	15.44	83.24	5.84	34.06	18.71	1.42
Prod. of gov serv	−1.51	−0.98	−10.43	7.66	−25.67	4.64	21.54	18.09	1.13

the Financial and Insurance Services industry, Sect. 4.5 discusses Transport, Storage and Communication, Sect. 4.6 analyses Real Estate and Other Business Services and Sect. 4.7 examines Producers of Government Services.

4.1 Hotel and Restaurant Industry: What Is the Future of Tourism?

February 2020, a cruise ship docked at a port in Jamaica with hundreds of tourists who sailed hundreds of miles to experience the "Jamaica no problem" culture. From the deck of the ship, they could glimpse in the distance a mural of Usain Bolt, the fastest man in the world and hear the sound of Bob Marley's "...Don't worry, about a thing... every little thing is gonna be alright..." Except, everything was not ok. News of the novel Coronavirus Disease (COVID-19) had surfaced. The World Health Organisation (WHO) declared the virus a pandemic. There was uncertainty everywhere. The local Ministry of Health barred everyone from leaving the ship out of fear that they might contaminate the island. The tourists, the local craft vendors, bar tenders, tour guides and other industry players were disappointed.

Most of the tourist come by air but cruise ships are highly anticipated by the locals who depend on the tourist spending to survive and support their families. On a visit to the island, each tourist spends roughly $134 US on average while the typical worker earns roughly $240 USD per week, 10% compared to $2,400 USD earned by their counterparts in the United States, or other developed countries. These wages are literally hand to mouth. Compensated less and therefore, should be less motivated. Jamaicans, nevertheless, maintain their hustling spirit and appear to be very happy, polite, and accommodating.[5] Traits they inherited from their foreparents, who were unpaid slaves and servants catering to generations of royalties, plantation owners, pirates, aristocrats, and bourgeoisie. Unlike their foreparents who were forced into slavery from Africa, many of these workers migrated from their rural and urban communities willingly to find employment in the tourism hot spots. Here they are accommodated by cotching with relatives or squatting in less than

[5] Jamaica ranked 37 on the Global Happiness Index in 2021 up from 60 in 2020. This is Jamaica highest rank on any index globally. A country birth out of struggles knows how to survive and increase the comparative in times of hardship.

hospitable circumstance characterised by poor physical and social infrastructure, that according to Hall and Jones (1999), can depress their morals and reduce their productivity.

The tourism industry is the largest foreign exchange earner for the island, bringing in over US$2 billion per annum. In 2019, the industry grossed as much as US$3.6 billion in foreign currency inflows, of which, hotels and restaurants were major contributors.[6] Over the past 20 years, output in the Hotel and Restaurant Industry has been gradually trending upwards from $25 billion in 2000 to $45 billion in 2019, see Fig. 4.3, Accommodation and Food Service GDP. Correspondingly expenditure in thesector grew steadily prior to the COVID-19 pandemic, see Fig. 4.4. Despite this, participants in the local hotel and restaurant industry have not enjoyed a reasonable standard of living from the income earned. They also have limited employee benefits and job security. Labour productivity in the industry has been trending downwards, declining from $450,000 per person in 2011 to roughly $395,216.78 JMD in 2019, see Fig. 4.5. While employment has trending upwards, increasing from 76,000 people to over 104,000 people over the same time frame, see Fig. 4.6. Over the last 5 years, employment growth in the industry has averaged more than 5% while productivity growth has been negative, see Fig. 4.7.

The government has a Tourism Enhancement Fund (TEF) where they offer benevolence to assist the sector, which is also a major employer of Jamaicans especially along the north coast. Productivity growth has slowed down markedly since the Global Financial Crisis (UN, 2017). According to the UN 2017, many factors have been proposed as underlying forces behind this phenomenon, including a slower pace of technological progress and the role of weak demand and lower capital investment. The COVID19 pandemic also hit Jamaica's tourism industry earnings severely, declining by 83% in 2020 along with a sharp decline in tourism expenditure.[7] There has been a reduction in global travel and tourism dependent businesses and workers. One Jamaica Union of Travellers Association (JUTC) operator sold his busses and diversified into construction equipment, trucks and tractors. In 2021, the industry is gradually rebounding but severe challenges persist including the efficiency of labour employed. This will be analysed in the next section.

[6] https://jis.gov.jm/jamaica-earned-us3-64-billion-from-tourism-in-2019-welcomed-4-3-million-visitors/.

[7] See Fig. 4.4.

Fig. 4.3 Accommodation and food service GDP (*Source* STATIN)

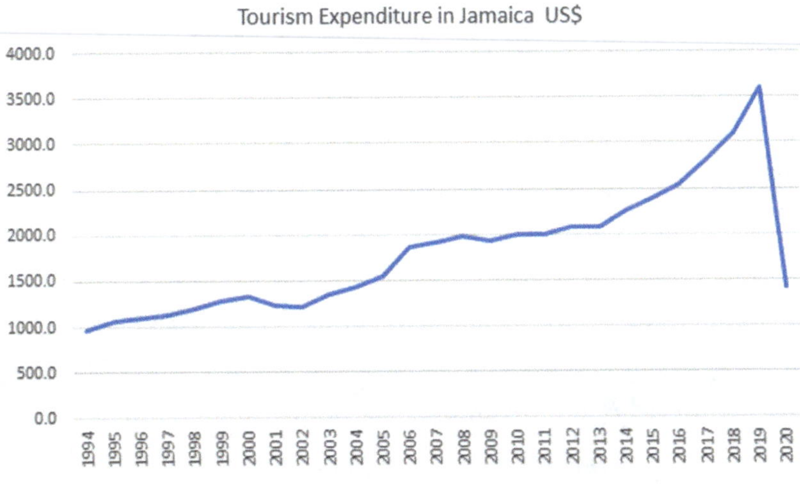

Fig. 4.4 Annual tourist spending in Jamaica (1994–2020) (*Source* STATIN)

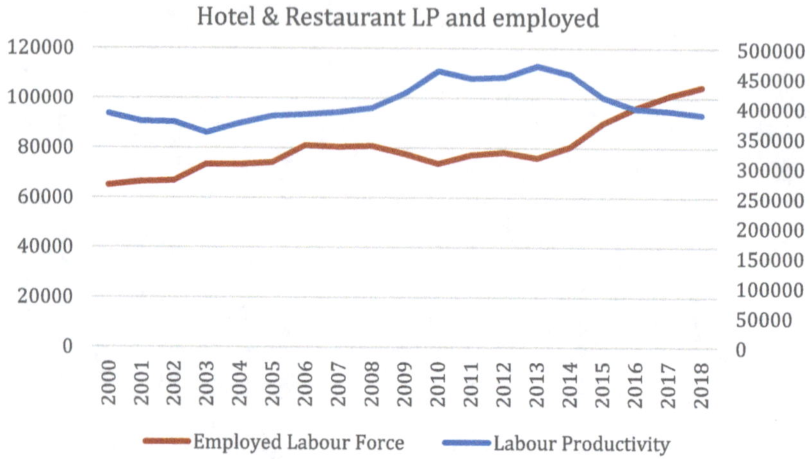

Fig. 4.5 Accommodation and food service employment and productivity (*Source* STATIN)

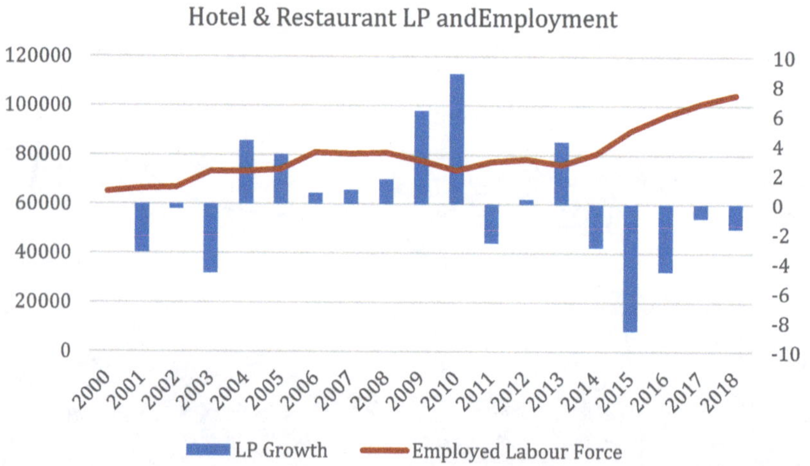

Fig. 4.6 Accommodation and food service productivity growth and employment (*Source* STATIN)

Fig. 4.7 Accommodation and food service: employment and productivity growth rates (*Source* STATIN)

4.1.1 The Optimal Level of Employment in the Hotel and Restaurants Sector

Employment in the Accommodation and Food Service industry has fluctuated between a low of 65,000 and a high 105,000 people over the last 20 years. Producing output per worker as low as $360,000 to a high of $470,000. As usual, the scatter plots shows the relationship between the number of person's employed and total output, and labour productivity, see Figs. 4.8 and 4.9 . The red dot on both curves shows the level of employment that produces the highest level of output, the orange dot shows the current level of output and productivity, while the green dot on both curves shows the level of employment that produces the highest level of productivity. Current level of employment is 104,300 people each with an average productivity of $389,000 JMD. At the highest level of productivity of $471,000 only 86,200 people would be required to produce the current output of $40.6 billion. At optimality, labour in this sector could fall by as much as 17% and still produce the highest potential output. If the current labour is employed at its highest level of productivity, the economy could produce up to $49.2 billion; about $8.6 billion or 21% more than it is currently producing now, see Table 4.2.

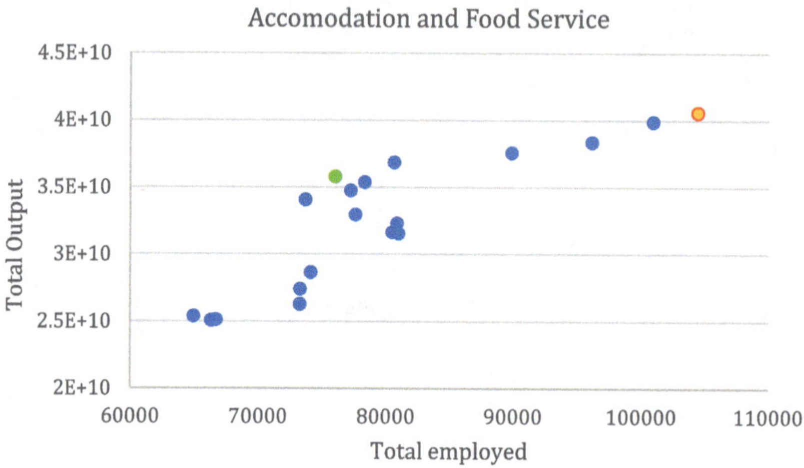

Fig. 4.8 Accommodation and food service output and employment scatterplot (*Source* STATIN)

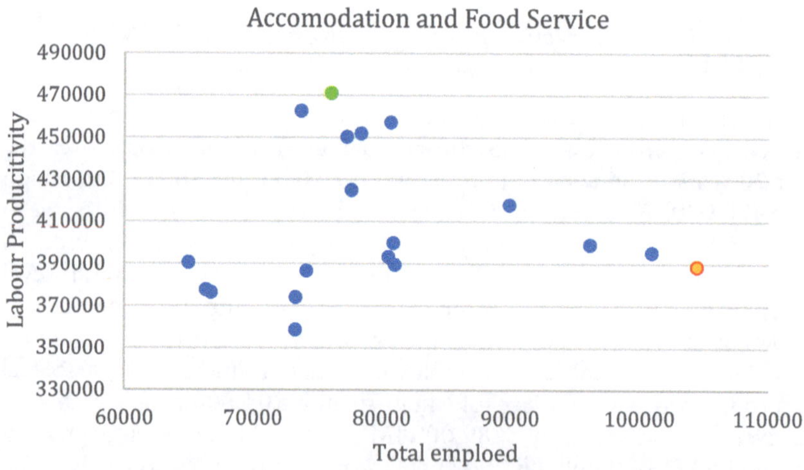

Fig. 4.9 Accommodation and food services productivity and employment scatterplot (*Source* STATIN)

Table 4.2 Labour usage monitor: accommodation and food services

Hotel and restaurant				
Labour usage monitor	*Current*	*Optimal*	*Excess*	*Percent*
Labour	104,375	86,156.94	18,218.06	17
Productivity	388,636	470,813	−82,177.79	−21
GDP	40,563,845,043	49,141,152,020.78	−8,577,306,977.53	−21

The industry currently employs 105,000 people and optimal output is achieved at 76,000 people therefore the industry is under employing 29,000 people.

4.2 Transforming Wholesale and Retail Trade with Innovate Entrepreneurship

The Wholesale and Retail Trade industry is characterized by entrepreneurs whose businesses are predominantly buying and selling of goods with little or no value added. It attracts roughly 60% of the people employed in informal economy in Jamaica (Torero et al., 2006). The Industry accounts for more than 18% of GDP, the largest share among all the industries. Wholesale and Retail Trade's contribution to GDP gradually increased in the early 2000's peaking at JM $1.25 billion in 2007 just before the Global Financial Crisis, after which it entered a downward trend to arrive at about JMD $1.17 billion in 2018. Johnston et al. (2000) noted that the Wholesale and Retail industry acts as a bridge between wholesaler and the retailer and is a major contributor to the productivity growth of Australia. In Jamaica however, labour productivity in this industry has declined from JMD $630 thousand in 2000 to $481 thousand in 2018 while employment has risen from 182 thousand people in 2000 to more than 244,000 people in 2018, see Fig. 4.11 and 4.12. This is like what occurred in the goods producing industries and must be investigated to see if labour is being over employed in the industry.

There are difference sources of productivity growth within and between countries. Van Ark et al. (2003) posits that technology infusion accounts for the heterogeneous productivity performance in the Wholesale and Retail industry between Europe and the United States. Alzyadat and Almusalamani (2021) found evidence to suggest that a 1% change in technology is associated with a 4.62% change in productive output of the Wholesale and Retail industry in Saudi Arabia. Johnston et al. (2000) noted that significant advantages can be derived for the Wholesale and Retail industry if labour saving technologies can be employed. This would require the facilitation of increased competition, changes in industrial relations legislation and factors specific to sub-industries. Australia managed to revolutionize their Wholesale and Retail Trade industry with the inclusion of productivity enhancing technologies such as bar coding and scanning as part of their inventory management which facilitated the development of a more productive electronic data system (Johnston et al., 2000). This helped to reduce the labour intensity in the Australian industry and improve total factor productivity. Hong Kong's situation is a bit different. Their increased productivity performance since 2005 was more labour intensive than capital intensive but with a more educated and trained labour force (Lee, 2012). Jamaica has been increasing the labour intensity but labour productivity and employment has been anticyclical, see Fig. 4.13. While Jamaica's productive process is labour intensive, they have not been able to harvest productivity gains because human capital is low, see Chapter 5. How labour is employed in Jamaica is suboptimal and needs dire attention if Jamaica is going to transform the productivity of the Wholesale and Retail Industry and improve the distribution of goods. The next section looks at labour usage in the industry (Fig. 4.10).

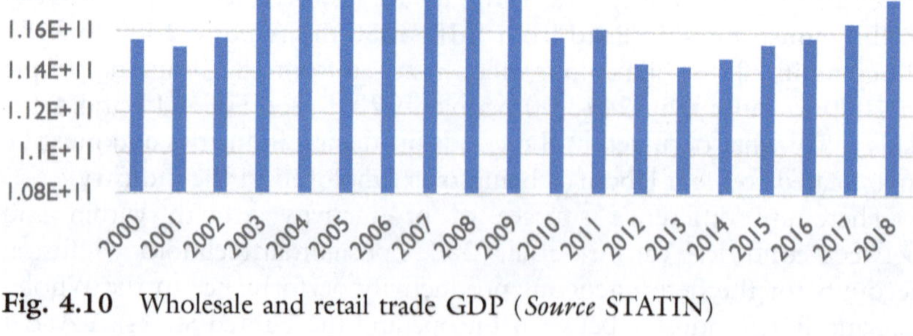

Fig. 4.10 Wholesale and retail trade GDP (*Source* STATIN)

Fig. 4.11 Wholesale and retail trade productivity and employment (*Source* STATIN)

Fig. 4.12 Wholesale and retail trade productivity growth and employment

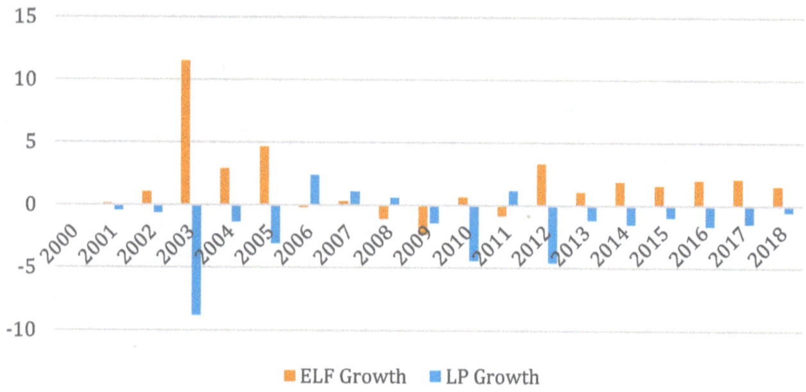

Fig. 4.13 Wholesale and retail trade employment and productivity growth rates (*Source* STATIN)

4.2.1 *The Optimal Level of Employment in the Wholesale and Retail Trade Industry*

The scatter plots show the relationship between the number of persons employed and total output, and labour productivity, see Figs. 4.14 and 4.15. The red dot on both curves shows the level of employment that produces the highest level of output, and the orange dot shows the current level of output and productivity, while the green dot on both curves shows the level of employment that produces the highest level of productivity. Current level of employment is 244,000 people each with an average productivity of $482,000 JMD. At the highest level of productivity of $633,000, only 186,000 people would be required to produce the current output of $118 billion. At optimality, labour in this sector could fall by as much as 24% and still produce the highest potential output. If the current labour is employed at its highest level of productivity, the economy could produce up to $154 billion, about 37 billion or 31% more than it is currently producing now, see Table 4.3.

4.2.2 *Improving Wholesale and Retail Trade in Jamaica*

Although this sector accounts for the greatest share of value-added GDP in Jamaica, greater value could be added with the inclusion of innovative entrepreneurship. This is the type of entrepreneurship that brings

Fig. 4.14 Wholesale and retail trade output and employment (*Source* STATIN)

Fig. 4.15 Wholesale and retail trade productivity and employment (*Source* STATIN)

Table 4.3 Wholesale and retail trade labour usage monitor

Wholesale and retail				
Labour usage monitor	*Current*	*Optimal*	*Excess*	*Per cent*
Labour	244,150	185,744.80	58,405.20	24
Productivity	481,843	633,352	−151,509.55	−31
GDP	117,641,855,361	154,632,913,120.29	−36,991,057,759.18	−31

forth new goods and services to the marketplace relative to buying a finished product and reselling it at a marked-up price for profit which currently exists. The emergence of the Fourth Industrial Revolution and the COVID-19 pandemic has posed a risk to the entrepreneurs that are engaging in this type of buying and selling which in its purest form can be seen as simply being traders. Not only are these entrepreneurs at risk, but so is the employment that they provide to the household and the spending the household contributes to aggregate demand in the Jamaican economy.

Over the last 4 years, anecdotal evidence would suggest that there has been a consistent rise in logistic related activity to support the purchasing of Wholesale and Retail Trade items in Jamaica and from overseas. More shipping companies are being opened or expanded, Shipme, Mailpack, MDcouriers and RocketShip, which are a few of the well-known couriers established to facilitate consumer purchase. There are also other smaller companies in the industry that are innovating for a share of the market. With competition, the Jamaican consumers of these services will be the main benefactors similar to the Digicel, Flow and Claro competition in the telecommunication industry a few years ago. There is also a move towards greater financial inclusion and deepening by the banking sector which has led to a more widespread provision of visa banking cards to assist with online purchases and payments. Most of the traditional banking debit cards in Jamaica only permitted access to ATM machines, but in recent times this has changed and now Jamaicans have access to at least one payment method to purchase from their favourite online stores. By connecting Jamaicans to the global economy with shipping companies to transport their items and the means to pay, in this manner, the traditional heavy reliance on local businesses in the Wholesale and Retail industry to provide certain goods is at risk. There are numerous examples spanning from the automobile industry where many people are importing their own vehicles from companies such as SBT Japan and B-Forward and clothing from stores such as Shein, Fashion Nova, Boohooman, and H&M just to name a few.

Many goods are not produced in Jamaica and attract customs duties, shipping cost plus mark up by the Wholesale and Retail businesses. Competition in the shipping industry was expected to reduce the cost to shipped items to Jamaica. However, back logs caused by the COVID-19 pandemic has increased shipping cost globally. In a low-income country like Jamaica, citizens will try to minimize their costs while at the same time procure goods with the highest quality and exactly what they desire. These citizens will sometimes source some goods cheaper online. Here they have a better chance of getting exactly the goods they want instead of a close substitute at a cheaper price than buying domestically. This will have implications for the Wholesale and Retail industry domestically and affect Jamaica's balance of payments. Additionally, many of the owners of firms in the Wholesale and Retail industry are operated by nationals of other countries including China who repatriate their profits or send them to offshore accounts. The most Jamaicans get from the industry is employment.

In order to increase competitiveness and overhaul low productivity growth in the Wholesale and Retail industry, improvements have to be made to incorporate better information about customers, faster information, smaller and more accurate inventories, sharp declines in operating margins and real consumer prices, increased firm and store size (McGuckin et al., 2005). These achievements will take time to achieve and will require a concerted effort.

4.3 Electricity Gas and Water Supply

The Electricity, Gas and Water Supply ($3,705,771.85) industry is the most productive in Jamaica. The industry's GDP that was on an upward trajectory before it was hit by the financial crisis in 2008 seems to have started to gradually recover towards the end of 2019, see Fig. 4.17. Electricity, Gas and Water Supply are utilities useful to every household, every business and the government as well. The industry employment has fluctuated between a low of 5,600 and a high 8,700 people over the last 20 years, see Fig. 4.18. The year 2016 saw a sharp decline in employment which coincided with a large increase in productivity growth for the same year, see Figs. 4.19 and 4.20. Jamaica's National Energy Policy guides the local energy sector towards the nation's vision 2030 goals.[8] The Public Sector Energy Efficiency and Conservation Program as well as the Energy Management and Efficiency Program encourages government entities to be more energy efficient. A 3D scan of Jamaica' oil and gas resources showed that Jamaica has 1.1 billion barrels of oil equivalent (0.067% of global reserves). The prospective partners did not follow through with the excavation after the end of the exploration. The government is currently seeking a partner to continue. Current oil price is approximately \$120 and the total value of Jamaica's deposit is roughly \$132 billion US or more than 8 times the country's GDP. In 2019, The Eight Rivers Energy Company completed a 115 MW renewable energy procurement project for Jamaica. The Government and the National Education Trust is in the process of equipping 30 schools with solar power to reduce their energy bills. The Petroleum Corporation of Jamaica (PCJ) is in the process of delivering solar powered energy solutions to remote homes. The project will begin with 45 off grid homes as potential beneficiaries. The Jamaica Public Service (JPS) in collaboration with the South Jamaica Power Company established a Liquefied Natural Gas (LNG) generation plant with the capacity to supply 300,000 households. Jamaica produces oil and gas-related products at the PetroJam refinery. In 2019, production of heavy fuel oil increased from 4.7 to 5.64 million barrels, E10 fell from 1.22 million barrels to 1.02, LPG 131,697 barrels to 123,089 barrels, Asphalt 155,076 to 128,613 and ADO from 976,479 to 760,217 barrels. Jamaica's total oil import bill decreased from 1.7 billion in 2018 to 1.6 billion in 2019. Jamaica imported US\$617.6 billion of crude oil in 2019, an increase of 19.6% from the year before. Crude oil comprises the bulk of Jamaica's import bill. Between 2018 and 2019, this paid for the 9.6 million barrels imported, a 28.4% increase on the previous year. Jamaica imported 22.1% less refined products in 2019 by PetroJam. Petroleum imported by bauxite companies also fell by 24.4% to \$158.1 million; marketing companies imported \$355 million. Jamaica consumed 20.8 million barrels of oil in 2018. The heat rate that shows the efficiency of energy production increased from 8.994/kWh in 2018 to 9.022 in 2019 (Fig. 4.16).

[8] ESSJ 2019.

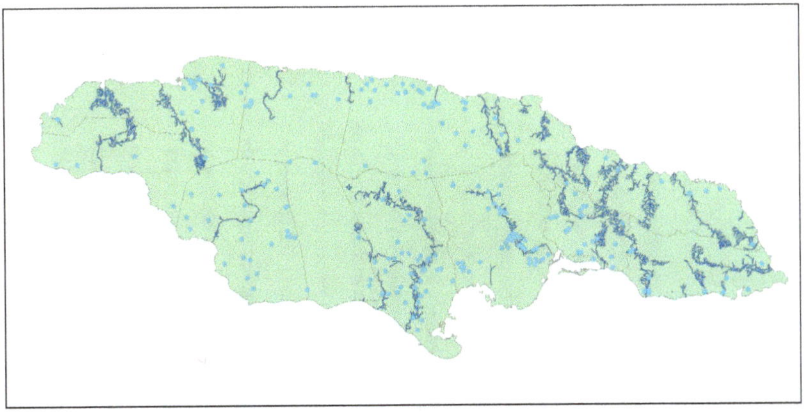

Fig. 4.16 Jamaica water resources (*Source* Department of Geography and Geology, UWI Mona)

Jamaica the land of wood and water. Figure 4.16 shows that Jamaica is well endowed with water across all 14 parishes except for Manchester. Nevertheless, the country has water challenges in rural areas which restricts agricultural output. In the urban areas, it disrupts the natural flow of business on a regular basis. As discussed in Sect. 3.2, there are challenges to effective water distribution on the island. These include, high levels of non-revenue water (NRW), high energy costs, vulnerability to the existing climate and climate variability, inadequate storage capacity in many parishes to increase the reliable yield and to ensure that there is sufficient supply of water during the dry season and a lack of infrastructure to move water from the northern draining basins to the southern agricultural villages.

4.3.1 Optimal Level of Employment in the Electricity, Gas and Water Supply Industry

As usual, the scatter plots shows the relationship between the number of persons employed and total output, and labour productivity, see Figs. 4.21 and 4.22. The red dot on both curves shows the level of employment that produces the highest level of output, and the orange dot shows the current level of output and productivity, while the green dot on both curves shows the level of employment that produces the highest

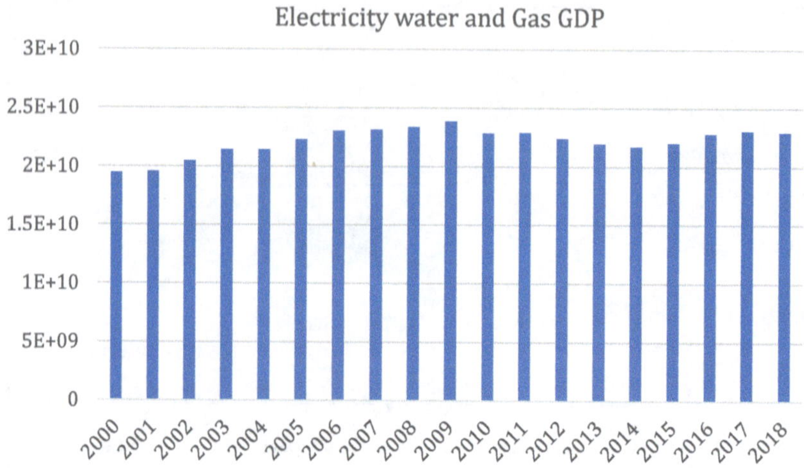

Fig. 4.17 Electricity and water supply GDP (*Source* STATIN)

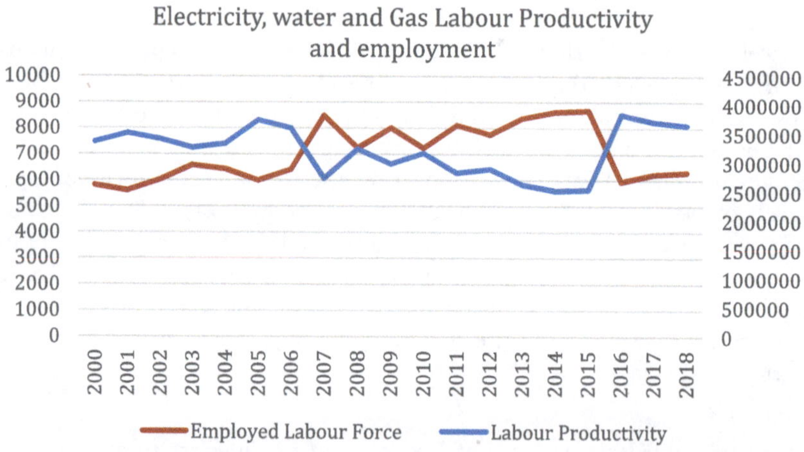

Fig. 4.18 Electricity, gas and water supply labour productivity and employment (*Source* STATIN)

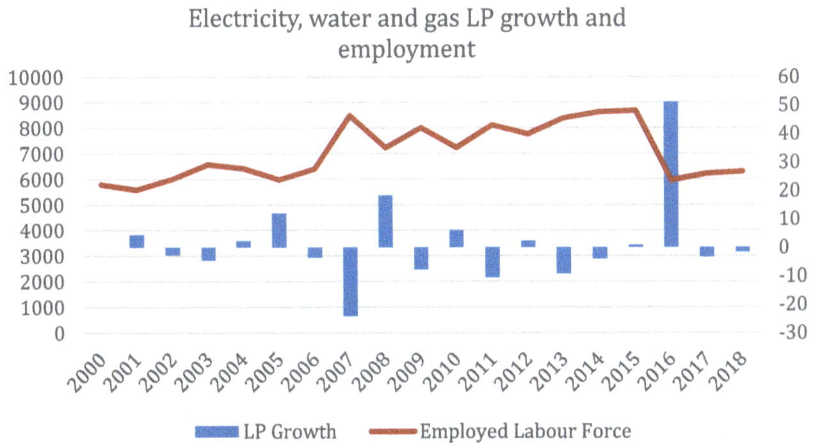

Fig. 4.19 Electricity, gas and water supply labour productivity growth and employment (*Source* STATIN)

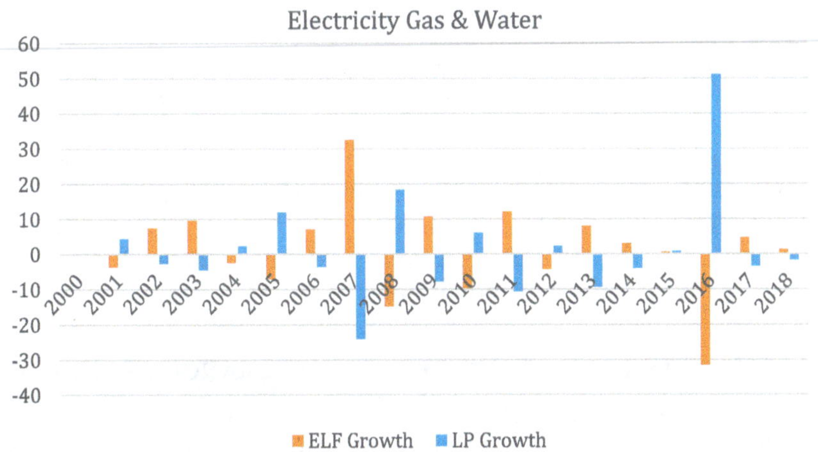

Fig. 4.20 Electricity, gas and water supply labour productivity and employment growth rates (*Source* STATIN)

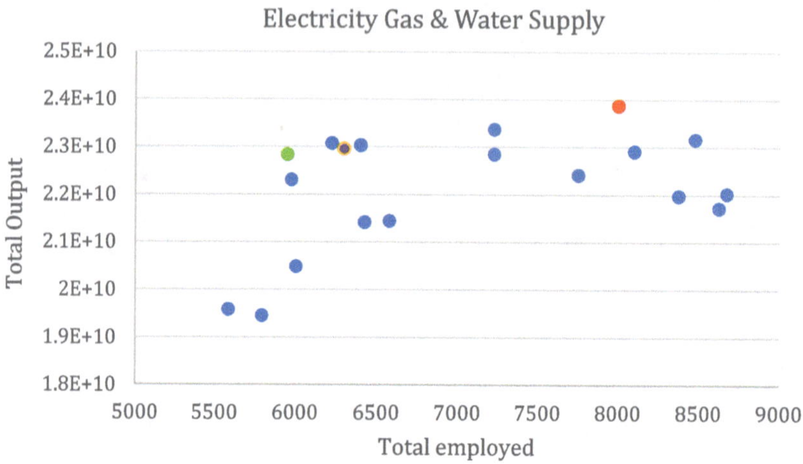

Fig. 4.21 Electricity, gas and water supply: output and employment scatter plot (*Source* STATIN)

level of productivity. Current level of employment is 6,300 people each with an average productivity of $3.64 million JMD. At the highest level of productivity of $3.84 million, only 5,990 people would be required to produce the current output of $23 billion. At optimality, labour in this sector could fall by as much as 5% and still produce the highest potential output. If the current labour is employed at its highest level of productivity, the economy could produce up to $240 billion, about 1 billion or 5% more than it is currently producing now, see Table 4.4.

4.4 Financial Services and Insurance Sector

Financial intermediation ($2,602,912.64) is the second most productive of all the industries in Jamaica. Over the last decade, there has been upsurge in financial and insurance products and firms. Bank of Jamaica policy has facilitated more financial market deepening, financial intermediation and insurance services. A total of 11 financial institutions are in Jamaica, 8 commercial banks, 2 building societies and 1 merchant bank with a total capitalization of $1.809 trillion. Growth in the industry has been influence by increase net interest income, fees and commissions on loans arising from an increase in the financial institutions loan portfolio. The BOJ's policy rate of 0.5% is the lowest in the nation's history but has

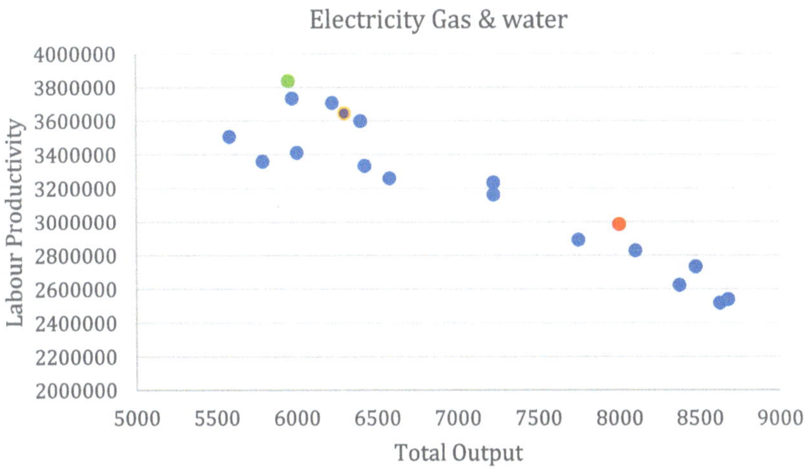

Fig. 4.22 Electricity, gas and water supply: productivity and output scatter plot (*Source* STATIN)

Table 4.4 Labour usuage monitor: electricity, gas and water supply

Elec gas and water				
Labour usage monitor	*Current*	*Optimal*	*Excess*	*Percent*
Labour	6,300	5,981.91	318.09	5
Productivity	3,643,179	3,836,903	−193,724.23	−5
GDP	22,952,028,885	24,172,491,562.44	−1,220,462,676.99	−5

been increased to over 4% as a response to the COVID-19 pandemic-induced inflation. Interest rate pass-through, though not perfect, has resulted in falling interest rates, making it cheaper for businesses and individuals to access loans. The demand and supply of loans increased. Between 2018 and 2019, the overall capital base increased by 8.1%, total deposits increased by 9.4% and total loans increased by 17.2%.[9] Total

[9] Check the 2019 ESSJ report produced by the Planning Institute of Jamaica for more details.

output in the Finance and Insurance Industry shows a slight increase from \$50.3 billion in 2000 to \$70.7 billion in 2019, see Fig. 4.23. Several legislative, regulatory and policy developments have contributed to the development of the sector in recent years. Parliament approved the Banking Services Rule in November 2019 that provides a framework for licensing and regulations of financial holdings under the banking services act.

Notwithstanding this, Labour productivity in the sector has been on the decline. Commercial Banks have reoriented their loan portfolio investment strategy; in 1990 more than 90% of the loans offered by these banks were for productive use, by 2019 it fell significantly to 50% of the loans for productive use. Table 4.5 shows the change in different categories of loans from the commercial bank from 1990 compared to 2019. Loans to the Household has increased by 39% while loans to Construction declined by 17%, loans to Manufacturing declined by 15%, loans to Agriculture declined by 7% and loans to Mining declined by 4%. Figure 4.24 also shows that 1989 91% of loans disbursed went to the productive use but has been reduced to 49% in 2019. Household loans are backed by collateral and seems to be a less risky loan option than any of the goods producing industries that have the worst output per worker in Jamaica among all the industries. The gradual increase in the number of consumer loans offered a corresponding decline in

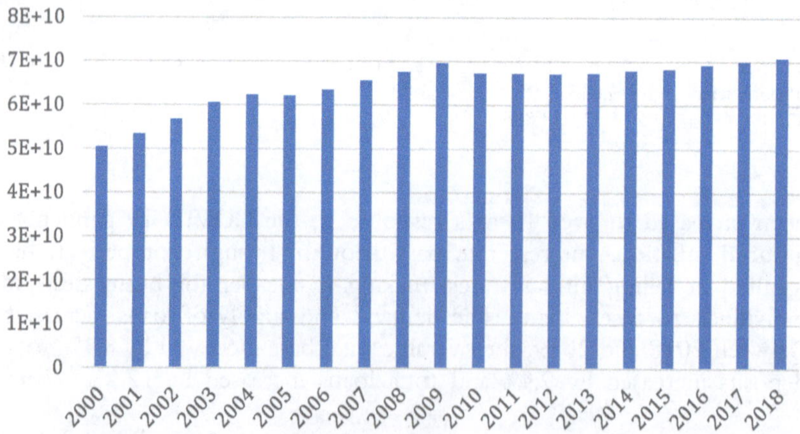

Fig. 4.23 Financial and insurance GDP (*Source* STATIN)

loans offered to the manufacturing sector and businesses, in general, see Fig. 4.1. Lack of funding has been a challenge for many emerging businesses, especially in the early stage when firms need it the most. According to Cusmano (2015), creating specialized platforms for SME public listing and providing financial resources for growth-oriented small and medium enterprises are necessary. The Private Sector Organization of Jamaica (PSOJ) and Balcostics Ltd (2013) conducted a survey of Jamaican SMEs and their studies highlighted the major challenges faced by SMEs that are accessing financing followed by government policies and administrative procedures.

This rerouting of capital might have been a contributing factor as to why the Jamaican economy is not growing as how it ought to. Small Island Developing States (SIDS) like Jamaica that are under-industrialized and have predominantly directed their attention and resources to low rewarding industries for the sake of job creation should try to re-engineer its strategies to achieve the desired long-run growth and developmental goals concurrently. The Grameen Banking system established by economist Muhammad Yunus to provide funding for rural area farmers in the Eastern Bengal is a good.

Table 4.5 Credit to the private sector by banks. ESSJ (1989 and 2019)

Bank loans to sector	*1989 (%)*	*2019 (%)*	*Variance (%)*
Personal (local residents)	8.9	48.7	39.8
Distribution	0.0	9.0	9.0
Tourism	8.2	5.8	−2.4
Professional and other services	7.3	6.5	−0.8
Construction and land development	23.2	5.9	−17.3
Overseas residents	0.0	6.7	6.7
Government services	7.8	1.6	−6.2
Manufacturing	19.6	3.9	−15.7
Electricity, gas and water	0.0	4.7	4.7
Transport, storage and communication	15.0	3.6	−11.4
Financial institutions	0.0	1.5	1.5
Agriculture	9.4	1.7	−7.7
Entertainment	0.0	0.3	0.3
Mining	4.0	0.1	−3.9
Food and beverage services (excluding tourism)	0.0	0.1	0.1

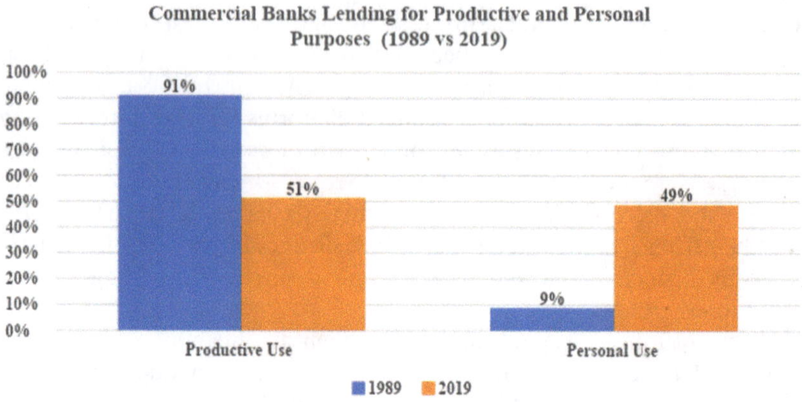

Fig. 4.24 Commercia banks' lending purpose (*Source* ESSJ [1989 and 2019])

The Grameen Banking system established by economist Muhammad Yunus provides funding for rural area farmers in the Eastern Bengal due to its structure, aims, objectives and target market which has been very successful at helping small businesses grow, sustain themselves and contribute to poverty alleviation. Since originating in 1976, more than 50% of Grameen borrowers have risen from absolute and relative poverty and reduced income inequality. More than 50 million people have risen from poverty from the exploitation of the Grameen Banking system. Having a diverse portfolio of investments will help to minimize risk and broaden the spread and reach to the rest of the world. Figure 4.25 shows the loan distribution structure for the Grameen Banking systems that has proven to reduce poverty in rural areas in Bangladesh. Although the aim of a commercial bank is profit making not poverty alleviation, understanding the dynamics of loan structure for productive capital can provide insight on the lucrative investment opportunities outside of the collateral backed household loans that might help to boost labour productivity in the sector. Jamaica's poor population is less than one million; it should not be difficult to employ a similar yet unique financing strategy to achieve the nation's growth and poverty alleviation objectives at the same time. Jamaica should seek to establish a legitimate funding source for MSME's similar to the Grameen Bank but tailored to suit the needs of the Jamaican economic environment.

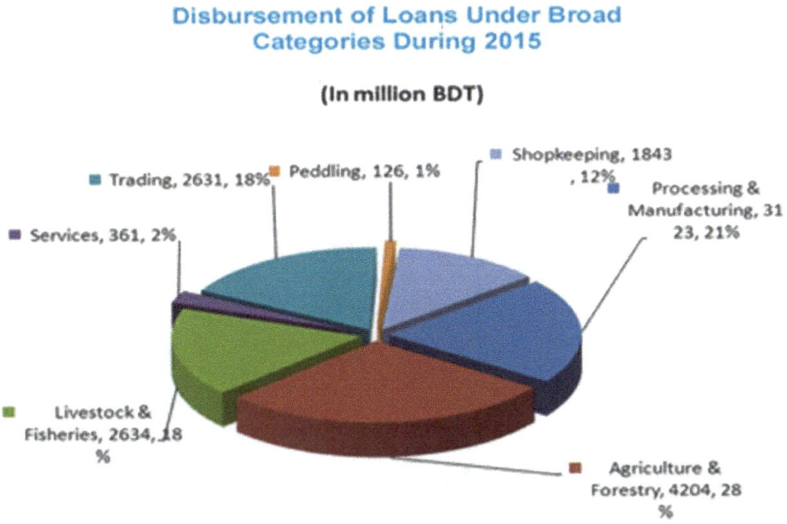

Fig. 4.25 Grameen bank distribution of loans

4.4.1 *The Optimal Level of Employment in the Finance and Insurance Sector*

Employment in the Finance and Insurance Sector shows a clear negative relationship with labour productivity. Employment increased from 27,000 in 2000 to 42,000 people by 2019. During this same time, labour productivity or output per worker declined from $4.8 million in 2004 to $2.6 million in 2019, see Figs. 4.26, 4.27 and 4.28. Scatter plots shows the relationship between the number of person's employed and total output, and labour productivity. The red dot on both curves shows the level of employment that produces the highest level of output, and the orange dot shows the current level of output and productivity, while the green dot on both curves shows the level of employment that produces the highest level of productivity, see Fig. 4.29. Current level of employment is 26,400 people each with an average productivity of $2.7 million JMD, see Fig. 4.30. At the highest level of productivity of $4.8 million, only 17,000 people would be required to produce the current output of $70 billion. At optimality, labour in this sector could fall by as much as 36% and still produce the highest potential output. If the current labour is employed at its highest level of productivity, the economy could produce up to $110 billion, about $39.5 billion or 56% more than it is currently producing now, see Table 4.6.

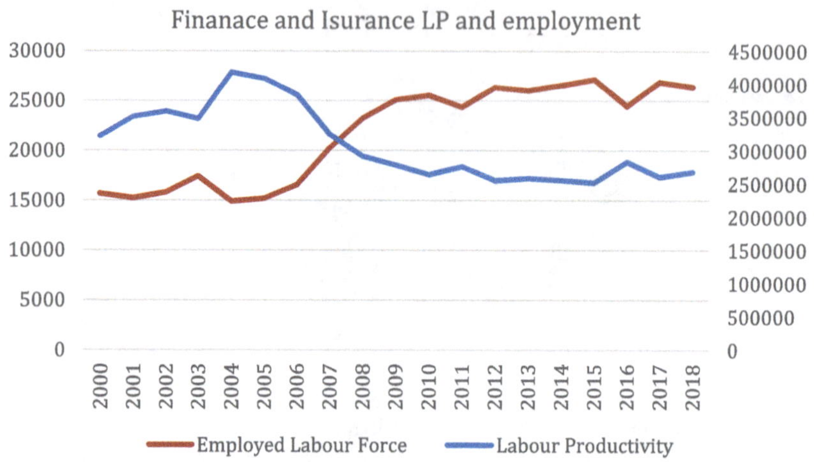

Fig. 4.26 Financial and insurance services employment and productivity

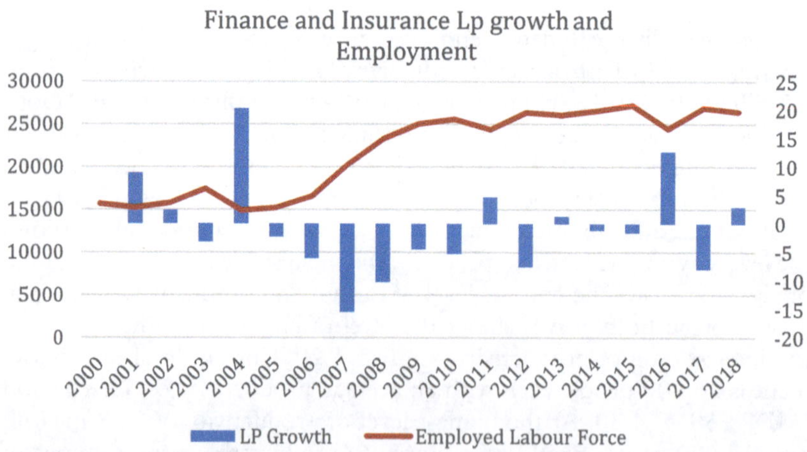

Fig. 4.27 Financial and insurance services productivity growth and employment (*Source* STATIN)

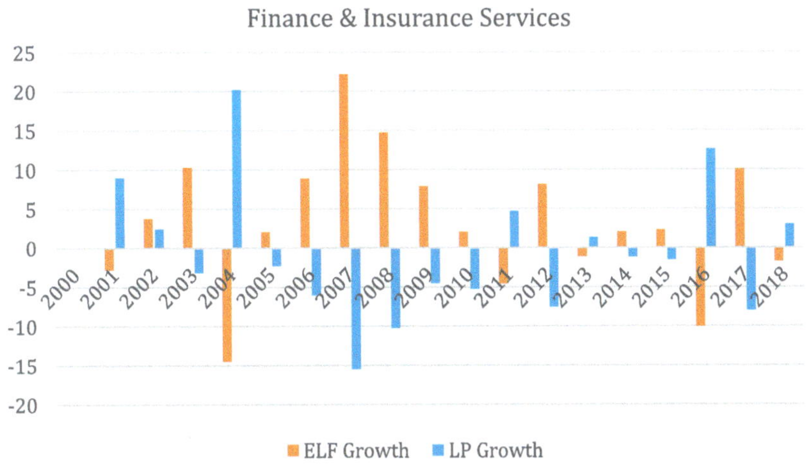

Fig. 4.28 Financial and insurance services employment productivity and growth rates (*Source* STATIN)

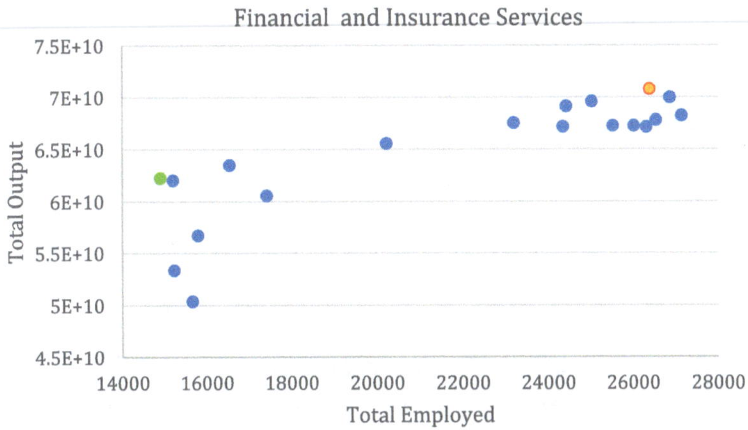

Fig. 4.29 Financial and insurance services: output and employment scatter plot (*Source* STATIN)

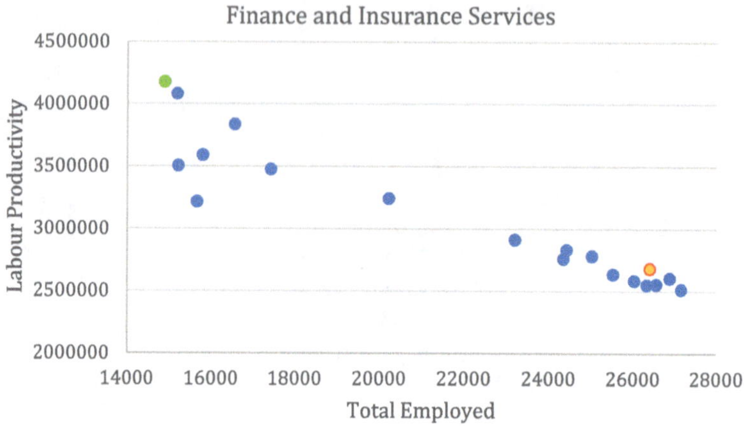

Fig. 4.30 Financial and insurance services: productivity and employment scatter plot (*Source* STATIN)

Table 4.6 Labour usage monitor: financial and insurance services

Finance insurance				
Labour usage monitor	*Current*	*Optimal*	*Excess*	*Percent*
Labour	26,400	16,938.14	9,461.86	36
Productivity	2,680,429	4,177,750	−1,497,321.40	−56
GDP	70,763,321,689	110,292,606,711.09	−39,529,285,022.42	−56

4.5 Transport, Storage and Communication a Push Towards ICT

Transport, Storage and Communication ($925,575.13), generate the 4th highest output per person on the island. Alam (2008) noted that productivity growth in services in the more advanced economies in Europe is largely driven by strong performances in industries such as transport, telecommunications and financial intermediation, services that have a greater propensity to use ICT. Van Ark et al. (2003) also noted that sectors that make use of ICT or are ICT intensive tend to have faster productivity growth. Jamaica too in recent years has been making a strong push towards increasing ICT's share of its service offering. There was liberalization of Jamaica's telecommunications industry in early 2000s which introduced Digicel as a competitor to Cable and Wireless Limited at the time. Digicel introduced per second billing which transformed Jamaica's telecom sector through a huge reduction in the cost of voice and data services.[10] Subsequently, according to the parliamentary report, millions of US dollars in FDI flowed into the island. It is estimated that Jamaica spends approximately $80.9 US million or 12% of total FDI flows on ICT improvement per annum. These included efforts to build out telecommunications network infrastructure, statistical instruments for mining and petro-chemicals, geological investigations, purchases of computers, data storage and network equipment. The ICT sector was poised to boost Jamaica's productivity and increase the value of output per worker. Alam (2008) indicated that service industries that draw on a more highly skilled workforce have shown higher productivity levels and simultaneously have grown at more rapid rates compared to other industries. To reduce unemployment and improve the growth prospects, the Government of Jamaica has identified Business Processes Outsourcing (BPO) as a useful strategy. There is great potential to enhance the value of output per worker in this service area and service overall on a macro level. Jamaica's ICT companies are mainly customer service driven and lower order services. India currently has the greatest share of the global BPO industry in which they offer bookkeeping and accounting, data entry and data processing, insurance processing and market research among other services in their BPO industry. Jamaica's BPO industry on the other hand is highly concentrated with low-skilled job requirements. If Jamaica can add new areas which require a greater level of thinking or higher-level skills, then there can be increased value added to the service producing sector (Fig. 4.31).

[10] http://www.japarliament.gov.jm/attachments/596_Information%20and%20Communications%20Technology%20(ICT)%20Policy.pdf.

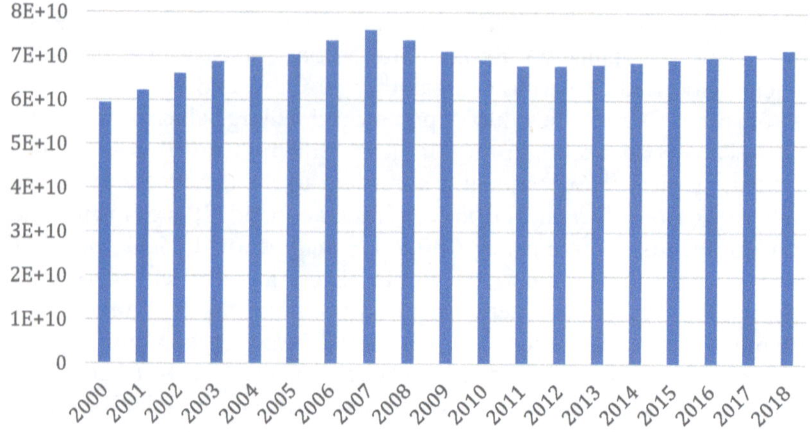

Fig. 4.31 Transport, storage and communication GDP (*Source* STATIN)

4.5.1 *The Optimal Level of Employment in the Transport and Communication Sector*

Employment in the Transport and Communication industry has fluctuated between a low of 66,000 and a high 82,000 people over the last 20 years, see Fig. 4.32, producing output per worker as low as $892,000 to a high of $983,000, see Fig. 4.33. As usual, the scatter plots show the relationship between the number of persons employed and total output, and labour productivity, see Figs. 4.35 and 4.36. The red dot on both curves shows the level of employment that produces the highest level of output, and the orange dot shows the current level of output and productivity, while the green dot on both curves shows the level of employment that produces the highest level of productivity. Current level of employment is 74,350 people each with an average productivity of $963,000 JMD. At the highest level of productivity of $983,000, only 73,000 people would be required to produce the current output of $71.6 billion. At optimality, labour in this sector could fall by as much as 36% and still produce the highest potential output. If the current labour is employed at its highest level of productivity, the economy could produce up to $73 billion, about $1.3 billion or 2% more than it is currently producing now, see Table 4.7 (Fig. 4.34).

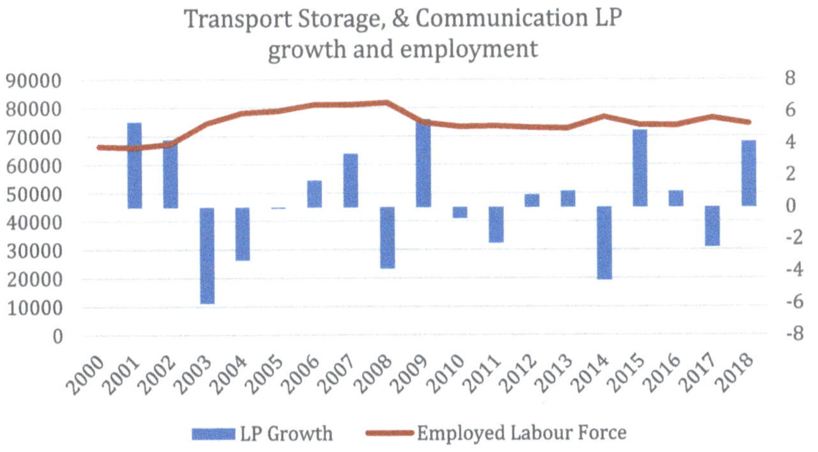

Fig. 4.32 Transport storage and communication productivity growth and employment (*Source* STATIN)

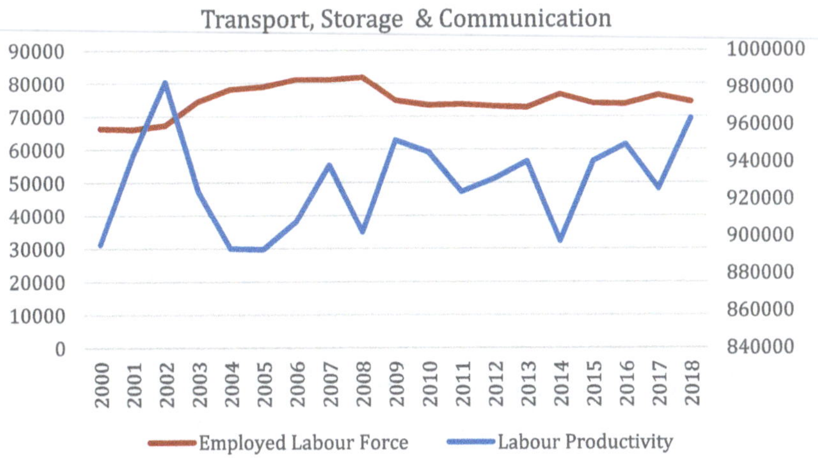

Fig. 4.33 Transport, storage and communication employment and productivity (*Source* STATIN)

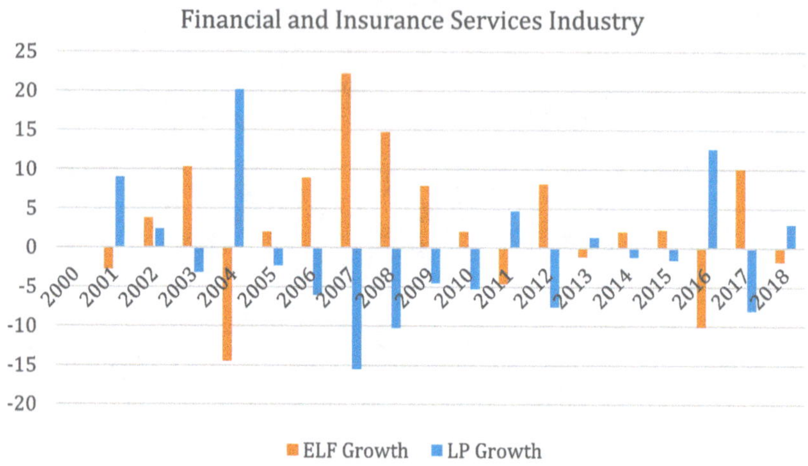

Fig. 4.34 Financial and insurance employment and productivity growth rates

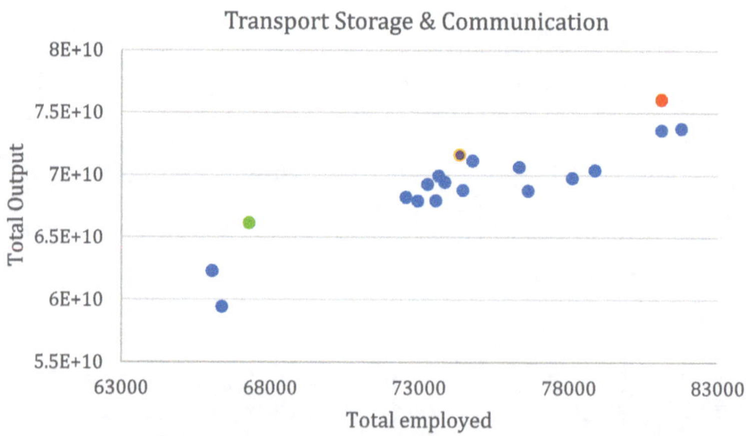

Fig. 4.35 Transport, storage and communication output and employment scatter plot (*Source* STATIN)

4.5.2 *Transport (Futuristic Rail Way Service to Deliver Goods)*

An efficient and reliable transport system is vital in the developmental process of any country. Jamaica's main transport company is the Jamaica Urban Transit Company (JUTC) which is supported by other local

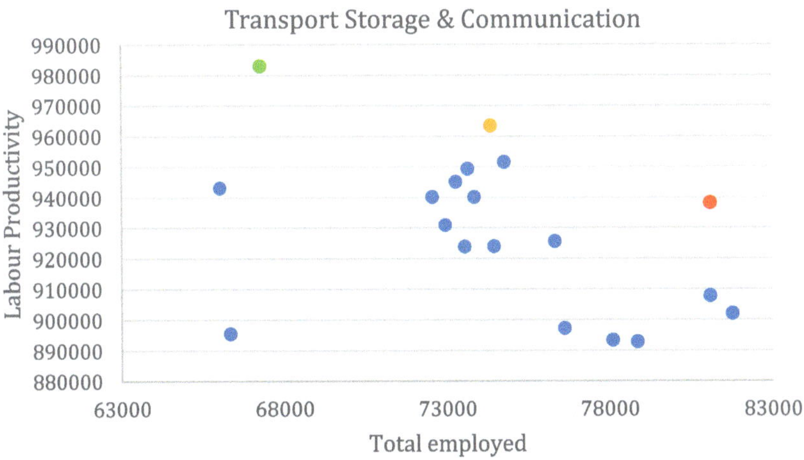

Fig. 4.36 Transport, storage and communication productivity and employment scatter (*Source* STATIN)

Table 4.7 Labour usuage monitor: transport storage and communication

Transport and storage				
Labour usage monitor	*Current*	*Optimal*	*Excess*	*Percent*
Labour	74,350	72,862.57	1,487.43	2
Productivity	963,285	982,950	−19,664.73	−2
GDP	71,620,266,212	73,082,339,237.98	−1,462,073,025.57	−2

services both public and private. Nevertheless, inefficiency exits in the transport industry and a railway service may help to reduce some of the inefficiencies which are likely contributors to low output per worker. These include fatigue from long travel hours and reduced hours worked due to punctuality issues. The Jamaica Railway Corporation (JRC) was enacted in 1960 with a mandate to transport both passengers and freights. The service has been suspended and restarted on more than one occasions which is an indication that there may be some nuances as it relates to management present. In Fig. 4.37, the red line represents the railway

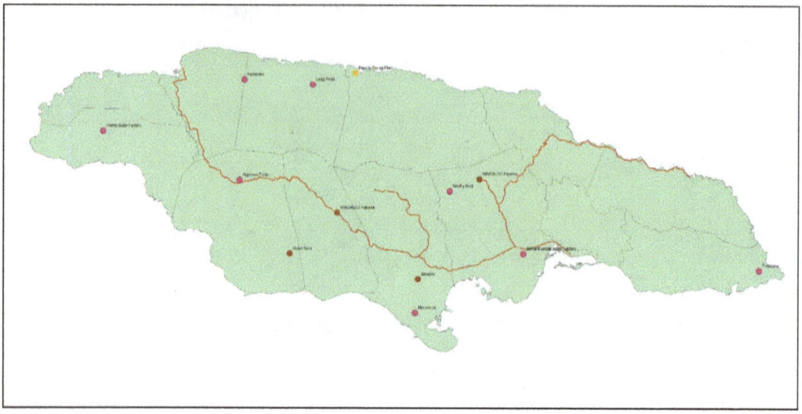

Fig. 4.37 Jamaica railway network (*Source* Department of Geography and Geology, UWI Mona)

network that flows between Montego Bay, St James, in the northwest of the island. The route travels south through St Elizabeth and then east through Manchester, Clarendon, St Catherine and to Kingston. Another section of the network extends from St Catherine to St. Mary and flowing along the north coast of Portland. This logistical network was instrumental in the movement of bauxite and major agricultural crops from farms and villages to markets, ports and docks. The pink dots are major sugarcane plantations while the brown dots are bauxite mines. The railways' network was smooth and efficient, but it has lost its importance in the Jamaican logistic network to more inefficient means of transporting goods. The railway industry has been hampered by short-sightedness and little to no investments or development of critical infrastructure that would see the railway system in Jamaica advance to twenty-first-century standard. In 1989 the Jamaica Railway Company (JRC) earned $49.29 million from the provision of freight passenger and mining services. This was 7.9% above the 45.68 million earned during 1988. Mining traffic from Alcan and Alcoa continued to be the dominant source of earning and accounted for 92.7% (45.67 million) of total earnings. Mining traffic usually accounts for approximately 85% of total revenue; however, the increase percentage realized in 1989 was a result of the closure of the Montego Bay line in May of that year and the consequent loss of freight and passenger revenue. However, at the end of the year the JRC lost the contract to ship traffic for Alcoa, and this had an extremely negative effect on the company's financial out-turn in 1990. To reduce the dependence on mining, JRC started a campaign by placing more emphasis on moving general cargo. Freight service was also increased from 2 to 5 days per week between Kingston and Montego Bay. As part of the rehabilitation of the service, industries such as tourism, mining, construction and petroleum may be some of the main benefactors.

The following project were to be studied to determine feasibility and if so be implemented:

i. The extension of railway tracks into the ports, via Kingston and Montego Bay, and conversion to container traffic to increase the railway's share of general cargo.
ii. The movement of passengers in Portmore St Catherine whose population is approximately 100,000 and is expected to increase significantly over the next three years when 10,000 houses will be constructed.
iii. The reopening of the Port Antonio line from Bog Walk to Buff Bay.

4.6 Real Estate and Other Business Services

The Real Estate industry in Jamaica has been growing gradually from $55.8 billion in 2000 to $64.3 billion in 2018, see Fig. 4.38. The upward trend is a response to increase in housing demand island wide arising from population growth and internal migration. More people see the need to own their own homes as well as tourist and expatriates want to live in Jamaica for various reasons. Despite the growth of the Jamaica Real Estate industry over the last two decades, the industry is still in its infancy stage based on Jamaica's developing country status. According to the draft national Housing Policy 2019, housing demand is roughly 15,000 units per year up to 2030, a 5000 unit increase on current production levels.[11] This plus the demand for commercial property will propel the Real Estate industry for years to come. The mass supply of residential housing solutions is produced by private contractors commissioned by the government while the private sector provides lower middle income housing.

[11] https://www.japarliament.gov.jm/index.php/publications/green-papers/2019-green-papers/2167-2019-green-paper-2-draft-national-housing-policy-2019.

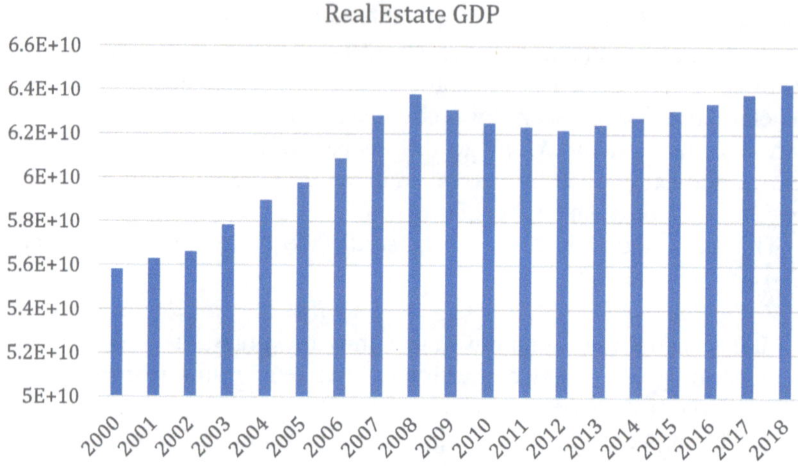

Fig. 4.38 Real estate and other business services GDP (*Source* STATIN)

Housing shortages across the island have resulted in increased price of homes especially in the Kingston and metropolitan area. A survey by the National Housing Trust on consumer housing budget indicated that roughly 17% of people surveyed can afford a house valued less than $4 million JMD, 20% can afford housing between $4 and $5 million JMD, 15% between $5 and $6 million and 16% can afford 8 million and above. From a supply point of view most of the current housing solutions are above 8 million, both governmental from the NHT and those from private contractors. The average 2-bedroom apartment in the KMA is going for more than $20 million JMD. It is mandatory for every employee to pay into the NHT to get cheaper loans to buy their houses. The NHT lends $6.5 million JMD the most and the remainder has to be sourced from the bank in the private market and many people do not qualify because they earn below the required threshold but are still expected to contribute to a trust that they will never benefit from. This put the housing market in the spotlight, and productivity in the Real Estate Industry has to be analysed.

4.6.1 *The Optimal Level of Employment in the Real Estate Sector*

Employment and productivity in the Real Estate industry shows a clear negative relationship. As employment increases from 37,640 in 2008 to 87,500 in 2019 more than twice as much, productivity in the industry declined from $1.43 million 2008 to $735,000 in 2019, see Fig. 4.39 almost halved. As usual, the scatter plots shows the relationship between the number of person's employed and total output, and labour productivity, see Figs. 4.42 and 4.43. The red dot on both curves shows the level of employment that produces the highest level of output, and the orange dot shows the current level of output and productivity, while the green dot on both curves shows the level of employment that produces the highest level of productivity. Current level of employment is 87,500 people each with an average productivity of $735,000 JMD. At the highest level of productivity of $1.48 million only 43,400 people would be required to produce the current output of $64.3 billion. At optimality, labour in this sector could fall by as much as 50% and still produce the highest potential output. If the current labour is employed at its highest level of productivity, the economy could produce up to $130 billion, about $64 billion or 102% more than it is currently producing now, see Table 4.8 (Figs. 4.40 and 4.41).

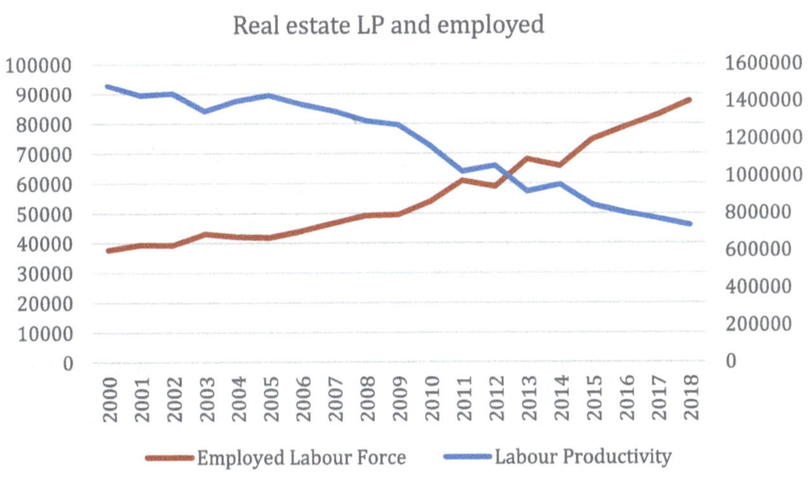

Fig. 4.39 Real estate and other business services employment and productivity (*Source* STATIN)

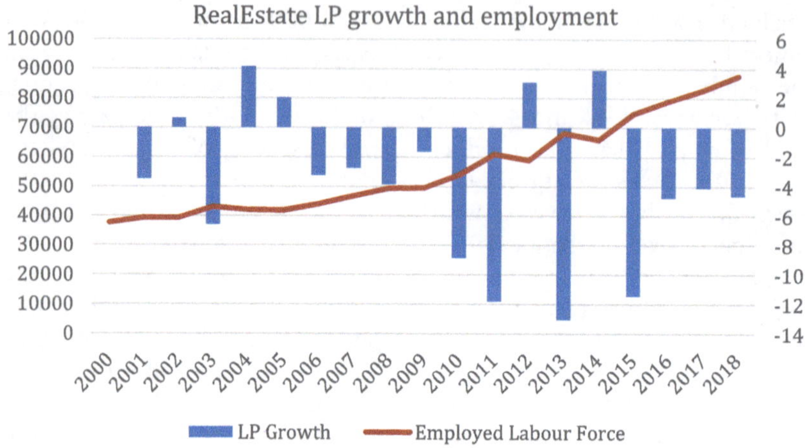

Fig. 4.40 Real estate and other business services productivity growth and employment (*Source* STATIN)

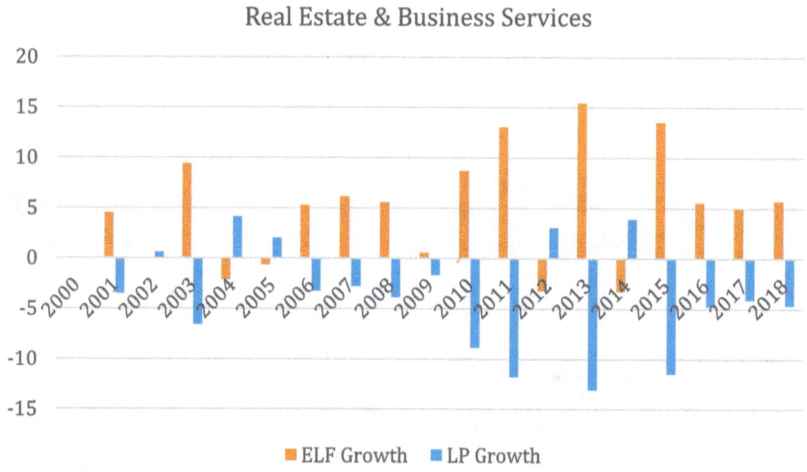

Fig. 4.41 Real estate and other business services employment and productivity growth rates (*Source* STATIN)

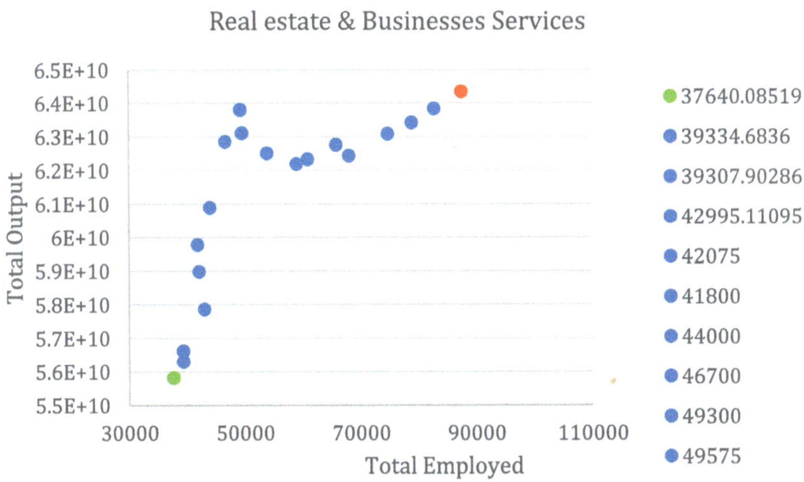

Fig. 4.42 Real estate and business services output and employment scatterplot (*Source* STATIN)

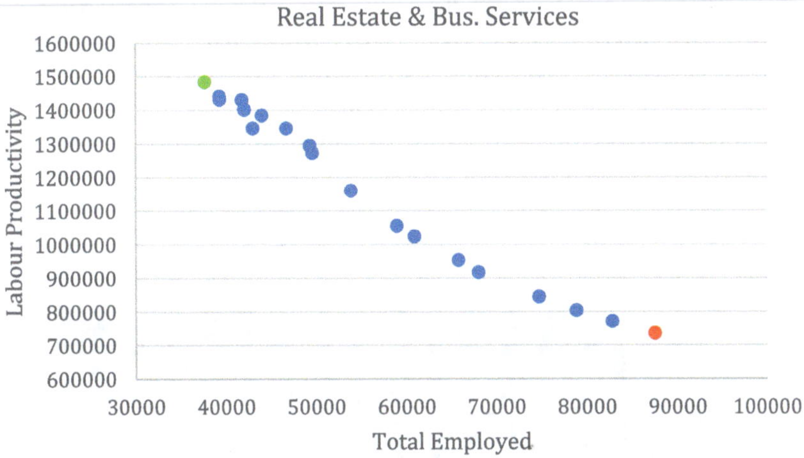

Fig. 4.43 Real estate and business services productivity and employment scatterplot (*Source* STATIN)

Table 4.8 Labour usage monitor: real estate and other business services

Real estate				
Labour usage monitor	*Current*	*Optimal*	*Excess*	*Percent*
Labour	87,500	43,391.13	44,108.87	50
Productivity	735,287	1,482,737	−747,449.83	−102
GDP	64,337,628,306	129,739,488,416.74	−65,401,860,110.44	−102

4.7 Producers of Government Services

Producers of Government Services Sector increased slightly from \$65.6 billion in 2000 to a high of \$69.4 billion in 2011 before falling steadily to \$68.9 billion in 2018 see Fig. 4.44. Political appointments play a real role in the labour supply of the government sector. Most Ministries, Departments and Agencies (MDA) hire people who they can trust not necessarily who can get the job done properly. The industry pays out over \$260 billion per year and has more than 300 pay scales which has become a focus of the government to curtail to reduce the pay mismatch for the same jobs across different industries within the public sector, in essence organizing and modernizing for the twenty-first century.

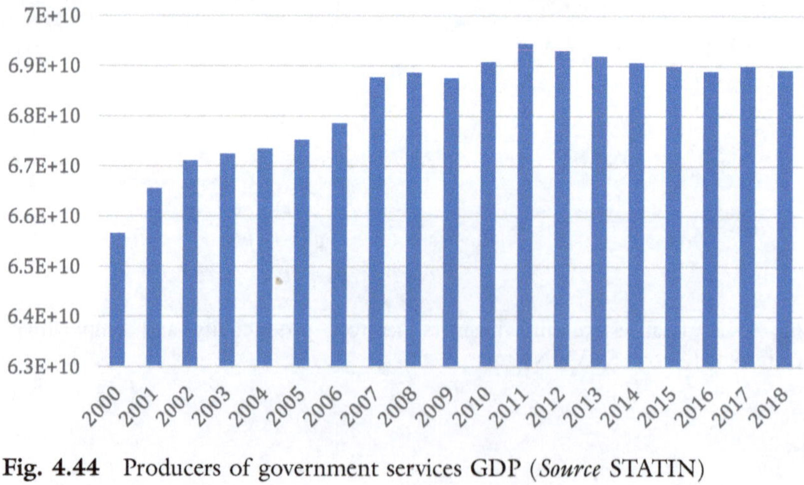

Fig. 4.44 Producers of government services GDP (*Source* STATIN)

4.7.1 *The Optimal Level of Employment in the Producer of Government Services*

The size of public sector has been a concern for many years. The international Monetary Fund (IMF) advises that Jamaica's public sector should be 9% of GDP employment in the Producers of Government Services industry that has fluctuated between a low of 114,000 and a high 167,000 people over the last 20 years, see Figs. 4.45 and 4.46, producing output per worker as low as $413,000 to as high as $580,000. Productivity growth as shown in Fig. 4.47 increases when employment growth falls within the industry. As usual, the scatter plots show the relationship between the number of persons employed and total output, and labour productivity, see Figs. 4.48 and 4.49. The red dot on both curves shows the level of employment that produces the highest level of output, and the orange dot shows the current level of output and productivity, while the green dot on both curves shows the level of employment that produces the highest level of productivity. Current level of employment is 167,000 people each with an average productivity of $413,000 JMD. At the highest level of productivity of $580,000, only 118,793 people would be required to produce the current output of $69 billion. At optimality, labour in this sector could fall by as much as 29% and still produce the highest potential output. If the current labour is employed at its highest level of productivity, this industry could produce up to $97 billion, about $28 billion or 40% more than it is currently producing now, see Table 4.9.

4.8 Discussion

Employment growth should increase output if employees are productive. Of the 1,222 million people employed in Jamaica, roughly 216,000 or 18% can be considered underemployed. These people could be employed elsewhere given that the industry could produce the same amount without them. About 59,000 or 5% points of this can become useful in other existing industries while the remaining 13% should be employed to new emerging industries if they wish to contribute positively to the island's productivity, see Table 4.10. From an industry point of view, 24% or 58,500 people from the Wholesale and Retail sector, 44,000 people or 50% of the Real Estate sector and 48,000 or 30% of government employees are underemployed and can be made redundant without

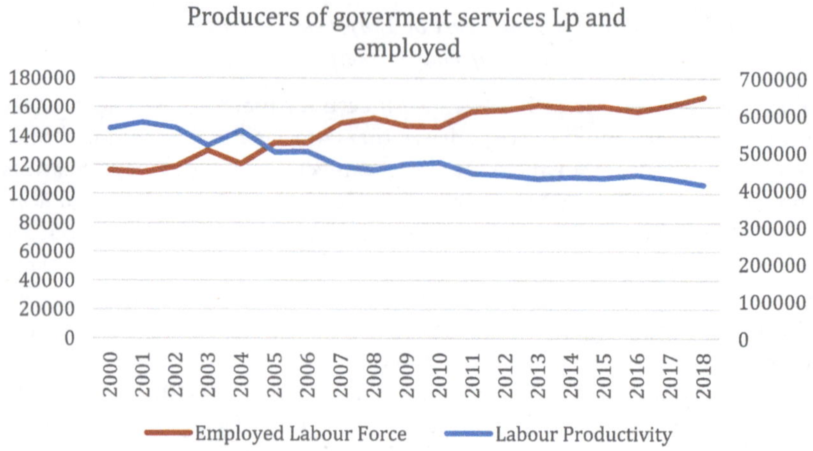

Fig. 4.45 Producers of government services productivity and employment (*Source* STATIN)

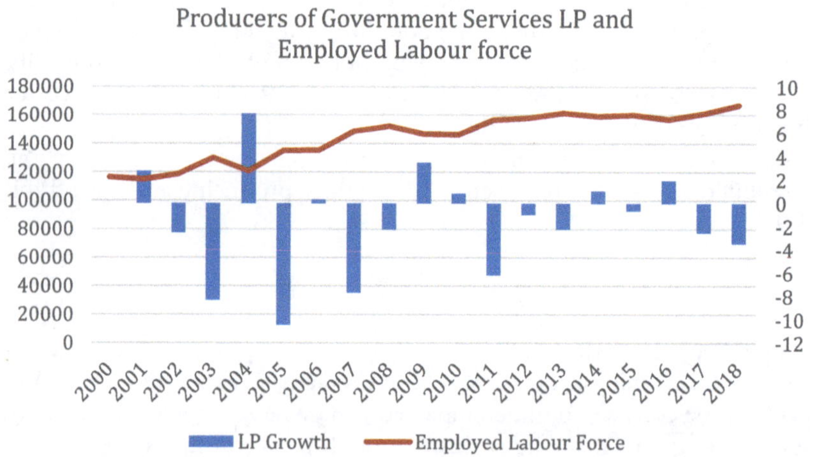

Fig. 4.46 Producers of government services employment and productivity growth (*Source* STATIN)

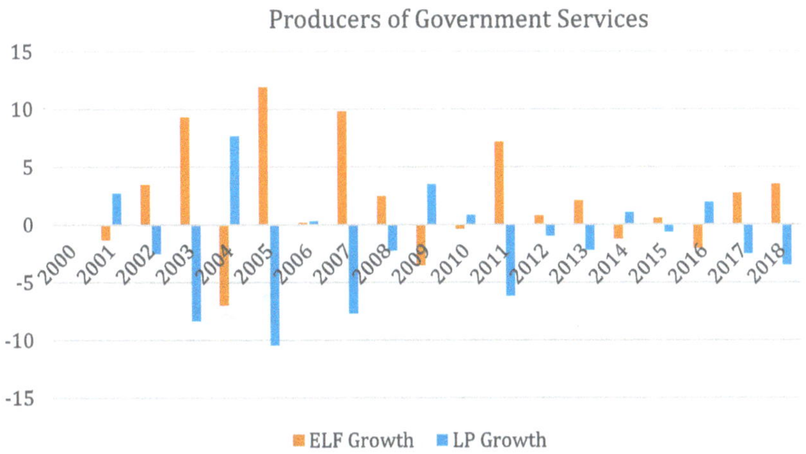

Fig. 4.47 Producers of government services employment and productivity growth rates (*Source* STATIN)

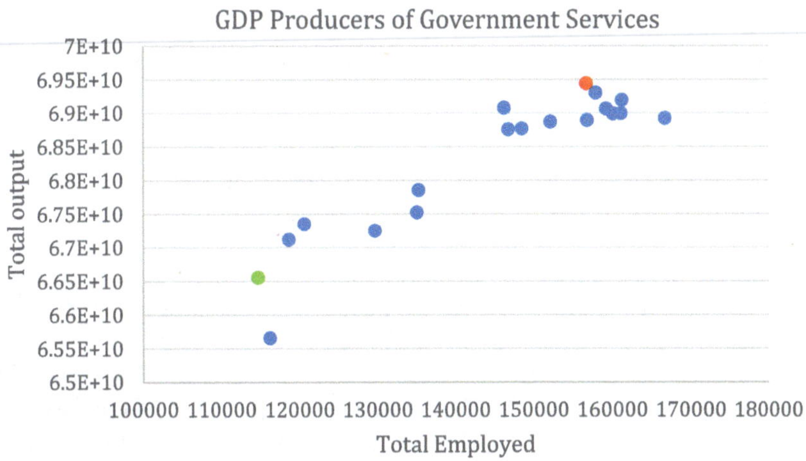

Fig. 4.48 Producers of government services output and employment scatter-plot (*Source* STATIN)

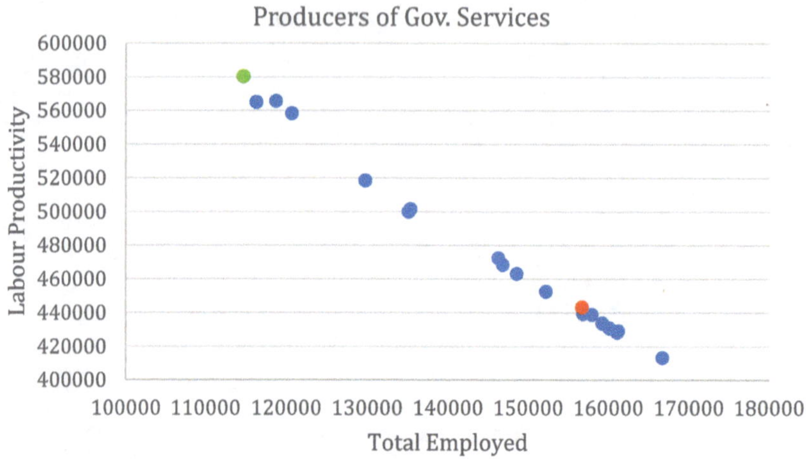

Fig. 4.49 Producers of government services productivity and employment scatterplot (*Source* STATIN)

Table 4.9 Labour usuage monitor: producers of government services

Prod. Gov. Services				
Labour usage monitor	*Current*	*Optimal*	*Excess*	*Percent*
Labour	166,725	118,792.19	−47,932.81	−29
Productivity	413,365	580,158	166,793.32	40
GDP	68,918,266,196	96,726,882,130.68	27,808,615,934.37	40

impacting the productivity levels of these sectors. Roughly 150,000 of the underemployed are from only three industries: Wholesale and Retail, Real Estate and Producers of Government Services. The agricultural industry is the only industry where workers are performing at their most productive levels and has no excess labour employed, see Table 4.10. Low productivity in this sector is caused by factors other than labour. It is imperative that employers get the most from their staff. In all the industries except agriculture, workers have been more productive in previous years and therefore have the potential to do better. Of the 79,000 people employed

Table 4.10 Labour usage monitor by industry

Industry	*Current labour*	*Optimal labour*	*Excess/under productive labour*	
Agriculture	194,300	194,300	–	0%
Manufacturing	79,100	65,417	(13,683.18)	−17%
Mining and quarrying	7,175	3,271	(3,903.56)	−54%
Construction	103,600	84,557	(19,043.31)	−18%
Whole sale and retail	244,150	185,745	(58,405.20)	−24%
Electricity, gas and water	6,300	5,982	(318.09)	−5%
Finance and insurance	26,400	16,938	(9,461.86)	−36%
Transp. St. Commun	74,350	72,863	(1,487.43)	−2%
Hotel and restaurant	104,375	86,157	(18,218.06)	−17%
Real estate	87,500	43,391	(44,108.87)	−50%
Prod. Government Services	166,725	118,792	(47,932.81)	−29%
Total labelled industries	1,093,975	877,413	(216,562.37)	−20%
Other	128,300	187,296	58,995.98	46%
Overall economy	1,222,275.00	1,064,708.61	(157,566.39)	−13%

in manufacturing, 17% or 13,700 of them could be employed elsewhere to increase their productivity. 3,900 or 54% of the 7,200 currently employed in the Mining and Quarrying sector, and 18% or 19,000 people from the Construction sector, see Table 4.10. Employees working in the Mining and Quarrying industry can be 119% more productive than they are now and therefore could see a 119% in GDP, and those working in real estate can be 102% more productive than they are now and could materialize the same increase in output. Finance and Insurance can be 56% more productive while Producers of Government Services can be 40% more productive; other industries can be 70% more productive than they are today, see Tables 4.11 and 4.12. These industries can all materialize the same increase in GDP.

4.9 Concluding Remarks

Small countries like Jamaica must get an understanding of how to employ labour if they wish to increase productivity. Most of the industries in Jamaica and many other SID's can be more productive than they currently are and therefore have excess underemployed labour that could be employed elsewhere. Government, entrepreneurs or investors must be able to decide when to hire or fire people. The efficient allocation of

Table 4.11 Productivity analysis by industry

Industry	*Current labour productivity*	*Optimal labour productivity*	*Excess/under labour productivity*	
Agriculture	279,111	279,111	-	0%
Manufacturing	660,311	798,428	138,116.68	21%
Mining and quarrying	2,244,554	4,922,804	2,678,249.92	119%
Construction	457,154	560,112	102,957.33	23%
Whole sale and retail	481,843	633,352	151,509.55	31%
Electricity, gas and water	3,643,179	3,836,903	193,724.23	5%
Finance and Insurance	2,680,429	4,177,750	1,497,321.40	56%
Transp. St. Commun	963,285	982,950	19,664.73	2%
Hotel and restaurant	388,636	470,813	82,177.79	21%
Real estate	735,287	1,482,737	747,449.83	102%
Prod. government services	413,365	580,158	166,793.32	40%
Total labelled industries	1,177,014	1,702,284	525,269.53	45%
Other	649,160	1,096,313	447,152.53	69%
Overall economy	527,854	605,971	78,117.00	15%

labour has been an obstacle for many developing countries outside of Asia especially in Africa, the Caribbean and Latin America. There are many other new and emerging industries, as well as old untapped industries, that if explored, might prove to be highly productive that can absorb this labour and increase the productivity of the country. Technology and innovation should be at the front line especially in agro-processing, big data, block chain and other modern innovations which can solve many of the labour efficiencies in many countries. The Corona Virus Disease (COVID-19) has made things worse and many small Island Developing States SIDs continue to struggle to fully understand and negotiate themselves upwards throughout these global economic bust and boom cycles. Industrial policy, if employed correctly, can increase the productivity and therefore the earning potential of a country.

Table 4.12 GDP output by industry

Industry	*Current GDP*	*Optimal GDP*	*Excess GDP*	
Agriculture	54,231,287,236.23	54,231,287,236.23	–	
Manufacturing	52,230,612,258	63,155,641,430	10,925,029,171.77	21%
Mining and quarrying	16,104,674,946	35,321,118,137	19,216,443,191.33	119%
Construction	47,361,172,954	58,027,552,484	10,666,379,530.26	23%
Wholesale and retail	117,641,855,361	154,632,913,120	36,991,057,759.18	31%
Electricity, gas and water	22,952,028,885	24,172,491,562	1,220,462,676.99	5%
Finance and insurance	70,763,321,689	110,292,606,711	39,529,285,022.42	56%
Transp. St. Commun	71,620,266,212	73,082,339,238	1,462,073,025.57	2%
Hotel and restaurant	40,563,845,043	49,141,152,021	8,577,306,977.53	21%
Real estate	64,337,628,306	129,739,488,417	65,401,860,110.44	102%
Prod. government services	68,918,266,196	96,726,882,131	27,808,615,934.37	40%
Total labelled industries	626,724,959,088	848,523,472,488	221,798,513,399.86	35%
Other	18,457,854,114	107,860,268,463	89,402,414,348.91	484%
Overall economy	645,182,813,201.60	740,663,204,025	95,480,390,823.40	15%

References

Alam, A. (2008). *Unleashing prosperity: Productivity growth in Eastern Europe and the Former Soviet Union*. World Bank Publications.

Alzyadat, J. A., & Almusalamani, M. S. (2021). The role of technological progress in the distribution sector: Evidence from saudi arabia wholesale and retail trade sector. *Journal of Distribution Science, 19*(3), 15–23.

Cusmano, L. (2015). *New approaches to SME and entrepreneurial financing: Broadening the range of instruments*. Organisation for Economic Co-operation and Development.

Hall, R. E., & Jones, C. I. (1999). Why do some countries produce so much more output per worker than others? *The Quarterly Journal of Economics, 114*(1), 83–116.

Johnston, A., Porter, D., Cobbold, T., & Dolamore, R. (2000). Productivity in Australia's wholesale and retail trade.

Lee, B. L. (2012). Output and productivity comparisons of the Singapore and Hong Kong wholesale and retail trade sectors, 2001–2008. *Asian-Pacific Economic Literature, 26*(2), 104–120.

Mark, J. A. (1982). Measuring productivity in service industries. *Monthly Labor Review, 105*(6), 3–8.

McGuckin, R. H., Spiegelman, M., & van Ark, B. (2005). The US advantage in retail and wholesale trade performance: how can Europe catch up? *Perspectives on a Global Economy.*

Rutkauskas, J., & Paulavičienė, E. (2005). Concept of productivity in service sector. *Engineering Economics, 43*(3), 35–41.

Torero, M., Robles, M., Hernández, M., De la Roca, J., Webber, M., & Thomas, D. (2006). *The informal sector in Jamaica*. IADB Economic and Sector Studies.

Van Ark, B., Inklaar, R., & McGuckin, R. H. (2003). ICT and productivity in Europe and the United States where do the differences come from? *Cesifo Economic Studies, 49*(3), 295–318.

United Nations. (2017, May). World economic situation and prospects: May 2017 Briefing. No. 102 https://www.un.org/development/desa/dpad/publication/world-economic-situation-and-prospects-may-2017-briefing-no-102/

CHAPTER 5

Human Capital, Productivity and Compensation in Jamaica

Education is one means through which individuals from the poorer echelons of society can acquire upward social mobility. Poorer people who acquire education put themselves in a good position to improve their productivity to earn higher incomes and increase their standard of living. From a national point of view, education is also the means through which a country adjusts its consciousness to know what is important on its path to development. In this chapter, we take a look first at Jamaica's Human Capital levels, then the link between human capital and productivity. Finally, we analyse the link between productivity and compensation.

5.1 Human Capital in Jamaica

Human capital (HC) or the potential of people's mental and physical effort is an important element of a country's productive process. According to Schultz (1961), investment in human capital leads to increase output.[1] In Jamaica, HC contributes positively to labour productivity with economic and statistical significance as shown by the results of Chapter 2. The higher the level of human capital, the higher Jamaica's

[1] https://la.utexas.edu/users/hcleaver/330T/350kPEESchultzInvestmentHumanCapital.pdf.

A. Haughton and W. Ivey, *Overcoming Productivity Challenges in Small Countries*, https://doi.org/10.1007/978-3-031-23301-2_5

level of productivity, the greater its GDP and the better its developmental pursuits. But what exactly comprises human capital and how can it be enhanced? In Chapter 2's analysis, human capital is measured by years of schooling (level of education) and compensation of employees (wages). This includes the economy's knowledge base as well as the expertise and skills acquired by people through training and education, (Marimuthu et al., 2009). Other factors like medical care expenditure, training, skills adaptation, ethics and punctuality all contribute to the efficiency of human capital development within countries (Becker, 1993).

Attitude towards life and work including levels of work discipline, punctuality and orderliness are attributes of good human capital (Myrdal, 1971). The level of word discipline is strongly linked to orderliness and are emphasized by the workers appreciation for their time as well as the time of others. Punctuality plays a huge role in efficient time management that is necessary to enhance human capital's contribution to productivity. Punctuality, however, is a big issue for many Jamaicans. Local events or meetings are held based on "Jamaican Time", a local colloquial term used to indicate that a one-hour grace period must be considered between the start time announced and the actual starting time of an event. Due to this expectation, people are late for most events and sometimes work. As a result, most things are either not done on time or not done properly so infrastructure depreciates quicker. This creates a backlog at the end of the day that needs be done another time, and the cycle continues. Jamaica's attitude to punctuality is a replica of the laid-back approach to business and commerce of the islanders, typical of the tourism destination. Tourist and FDI visitors comment that the island is on a sort of "go-slow" which reduces output per unit of time and is reflected in the low labour productivity. Somewhat the opposite of the anxious, fast-paced, early and on-time approach to business is used to facilitate increased productivity in modern metropoles.

Over the past 20 years, Jamaica's level of human capital has not changed relative to other countries around the world. Figure 5.1 illustrating Jamaica's human capital index relative to other Caribbean countries. In comparison to its Caribbean counterparts, Jamaica's score on the global Human Capital Index was only greater than that of the Dominican Republic & Haiti in 1990. The Dominican Republic, however, managed to surpass Jamaica in 2015. Jamaica's score of 2.59 in 2017 is less than that of Barbados, Belize, Dominican Republic and Trinidad Tobago, only better than Haiti, which has extenuating circumstances. This is a cause for concern. What factors contribute to Jamaica's poor human capital development relative to other Caribbean countries? What can be done to improve this level of human capital to boost productivity and output?

From a firm's point of view, there are 4 attributes that are important to human capital development (Garavan et al., 2001), 1. flexibility and adaptation, 2. enhancement of individual's competence, 3. their contribution to organizational efficiency and 4. the employability of the individual. Firstly, individuals must be flexible and able to adapt easily to new tasks or different work environment and conditions. This concept has been challenging for Jamaicans to grasp. The country sometimes approaches production and business in a traditional rigid way that does not allow much room for on-the-job flexibility. The situation is now exacerbated given that the firms' human capital needs have been evolving due to the COVID-19 pandemic. As the global economy transitions from the industrial era to the digital era, just as it did from the agrarian era to the industrial era in the middle of the twentieth century, some labour will become redundant and must evolve to survive. Robotics, AI, big data and machine learning and the 3rd internet which allows value to be transferred, require employees to be flexible and adapt to the new

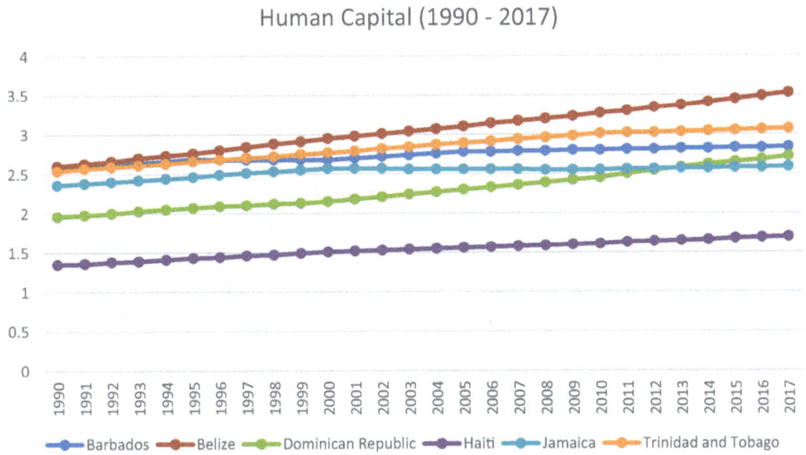

Fig. 5.1 Human capital in the Caribbean (1990–2017) (*Source* Penn World Tables)

work environment to fill these needs that can increase their employability. Training and education will be needed to get labour to the level that is required to participate in these new productive processes. Seminars and workshops are also essential to enhance individuals' competence. This along with mentorship can increase the employability of the individual and enhance their ability to contribute to the efficiency of the organization.

The aim is really to improve human development to increase the standard of living of a country's citizens which cannot be done without sustainable increases in human capital. Along with the stagnation in human capital, Jamaica's Human Development Index (HDI) has not changed over the past 10 years, which is an indication that the standard of living of Jamaicans has not really improved either. Human capital is the foundation of human development and knowledge acquisition is the root of Human Capital. The next section explores the level of education and training of Jamaica's workforce.

5.1.1 *Years of Schooling, Training and Education*

Jamaica and many other SIDs have realized low levels of output per worker and low growth in output partly due to a lack of proper formal educational attainment by a significant portion of its labour force. Higher education and training are important to equip people with the expertise needed to do well and improve at any job. Average years of schooling in Jamaica is about 9.7 years where females stay in school longer than males, 10.2 years relative to 9.3, respectively.[2] This is 4 years less when compared to 13.4 on average in the United States and Canada for example. More years of schooling means a more educated workforce. The benefit of education to productivity is that it allows workers to use physical capital more efficiently and it drives the development and diffusion of new technologies (Sweetman, 2002). Furthermore, education produces human capital which affects all levels of the economy and is a crucial determinant of productivity growth in both the medium and long terms; it is a key element of working smarter (Sweetman, 2002).

Data shows that as late as 2019, approximately sixty percent (60%) of people employed in Jamaica have no educational qualifications, see

[2] http://hdr.undp.org/en/countries/profiles/jam.

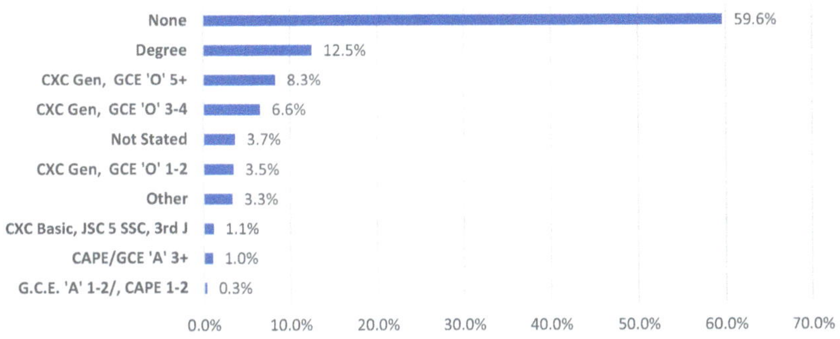

Fig. 5.2 Employed labour force by highest exam passed (2018–2019) (*Source* STATIN)

Fig. 5.2.[3] This makes it difficult for them to impact their individual, firm and national productivity levels. Another twenty percent (20%) have pre-university qualifications while only thirteen percent (13%) have university-level education. Pre-university education may suffice for the performance of basic tasks, simple production of goods and services, or even to equip workers with the potential to utilize technology in the workplace. However, university-level education is more likely to produce workers that are endowed with the capacity to be innovative or develop new technologies that will help Jamaica converge faster towards a knowledge-based economy.

Maneejuk and Yamaka (2021) analysed the impact of higher education on economic growth in the Association of Southeast Asian Nations five (ASEAN-5) countries. They found that higher education impacts become twice as strong when enrolment rates are greater than 49.5% in Thailand and 64.1% in Singapore. Though the other countries (Malaysia and Indonesia) in the study did not have a non-linear effect, the sign of tertiary education was steadily positive. Maneejuk and Yamaka (2021) concluded that secondary enrolment rates positively affect economic growth, but higher education is the key to future growth and sustainability. Jamaica's secondary enrolment rate of 87% is higher than the calculated threshold for the ASEAN-5 countries with a non-linear effect involved in this study. However, the Asian countries' economies have

[3] Data from the Statistical Institute of Jamaica (STATIN).

progressed at a more rapid pace than Jamaica. The ASEAN-5 countries were on a path to become the world's fourth-largest economy by 2030 (Australian Gov., 2017), while Jamaica's economy has been barely growing, although demonstrating positive signs from its turnaround from its high public debt and low GDP growth before the outbreak of the COVID-19 pandemic. The difference in futuristic growth pursuits is reflected in the difference in tertiary enrolment rate between Jamaica and the Asian 5 countries. Jamaica's average tertiary enrolment rate of 22% is less than half of the threshold (49.5 and 64.2) for higher education enrolment that is deemed necessary to accelerate economic growth as estimated by Maneejuk and Yamaka (2021).

Based on the type of job, training is needed to enhance the individual's competence in the workplace. Becker (1993) found evidence to suggest that better trained workforce contributes positively to productivity in countries. A decomposition of the labour force by the training received reveals that approximately 65.8% of the labour force in Jamaica have not received any training, see Fig. 5.3. This is similar to the ratio between the highest exam passed by people in the workforce. The low training and education phenomenon have contributed to the erosion of Jamaica's competitiveness in the global market space as outlined by the Global Competitiveness Report. In this report, competitiveness is described as the set of institutions, policies and factors that determine the level of productivity within an economy. The data shows that Jamaica has slipped 5 places from a rank of 75th in 2016 to 80th as of 2019 out of 140 economies assessed.

To increase years of schooling and training levels, the Government of Jamaica (GoJ) in 2021 made it mandatory for high school students to do a compulsory 7 years instead of the usual five years and the optional two. Instead of Grade 7–11, 6th form (grades 12 and 13) is now mandatory. This seemingly is to solve the low years of schooling issue but has sparked a local debate on the country's capacity to facilitate the change. Many schools are already overpopulated and cannot fully attend to its current student cohort. As a result, many students fall through the cracks and complete high school without passing any subjects to move to higher education. Many schools that already have 6th form can only accommodate a small number of the student cohort who graduate from grade 11 (5th form) but not everyone. Currently, approximately only 35% of the 40,000 students completing grade 11 enrol in a 6th form programme. Some students as outlined earlier are not inclined to education, and

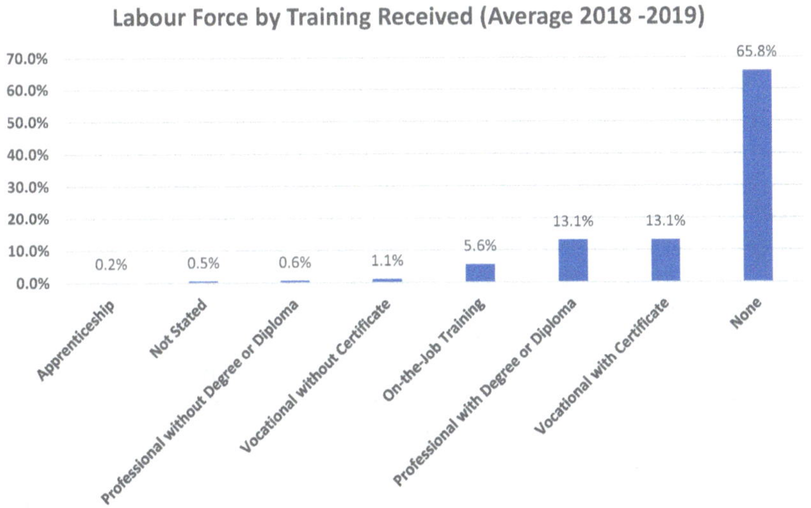

Fig. 5.3 Labour force by training received (*Source* STATIN)

some are inclined to training for vocational jobs. The HEART TRUST NSTA is the national vocational training institution that facilitate skills learning across Jamaica. They also provide working tradesmen with on-the-job certification based on an assessment that they are competent in their profession. More public sensitization will be required to allow more tradesmen to know of this avenue to become certified in Jamaica. This way, they have qualifications and can negotiate better paid wages and contracts.

The quality of the local economy plays a critical role in labour market outcomes and is also a major determinant of per capita economic growth (Sweetman, 2002). The economy must be able to absorb the labour that it educates or trains and compensate them accordingly. If the economy is populated by lower-level jobs where the educational demand is low, then there is less need for a highly educated population. Those with higher education are sometimes told that they are overqualified for some jobs they apply for. Many Jamaicans who aspire to earn high incomes to escape poverty and elevate from the lower to the middle or upper class are under the impression that this can only be done through migration, entrepreneurship or illicit activities. Very few see education as the way out in Jamaica. Many migrate to apply their trade elsewhere because education has proven to assist people to achieve upward social mobility in North America. Fogel (1999),[4] for example, points out that at the beginning of the twentieth century only 10% of Americans and Canadians were above the poverty level based on their real incomes. Fast forward to the twenty-first century, the figures have reversed, 86.3% of the people or 6 out of every 7 individuals are now above the poverty line in the United States, and 88.9% or 8 out of every 9 individuals are above the poverty line in Canada.[5] According to Fogel (1999), the reversal in poverty statistics in the United States and Canada is a direct result of productivity improvement driven mainly by increases in education attainment of the workforce. Fogel (1999) noted that in these countries, even the poor are rich when compared to the economic conditions of the past or in comparison to the standards in developing countries presently.

A lack of basic education and low levels of training by the majority of a country's workforce might trap them in an unproductive low-income spiral with low associating wages. In Jamaica, upward social mobility from productivity gains become an arduous process due to the low levels of training and education which has been reflected in the productivity out-turns at the macroeconomic level. Many of these uneducated and unskilled earn minimum wage or close and find it difficult to survive. The impact of the COVID-19 pandemic may make the situation even worse.

5.1.2 The Impact of COVID-19 on Human Capital Development

The COVID-19 pandemic sent a shockwave throughout the education systems in many countries across the globe which is likely to have a ripple effect on human capital development, future output and the earning potential of the cohort of students in school today. More than 168 million school children globally missed out on learning in class, as schools in many countries remained largely closed for almost an entire year due to coronavirus disease-related lockdowns.[6]

[4] Fogel, R. W. (1999). Catching up with the economy. *American Economic Review, 89*(1), 1–21.

[5] Wheaton, L., Giannarelli, L., & Dehry, I. (2021). 2021 Poverty Projections.

[6] United Nations (March, 2021). Over 168 million children miss nearly a year of schooling, UNICEF says https://news.un.org/en/story/2021/03/1086232.

In Jamaica, many students have been disengaged from learning as schools have had little or no regular contact with many students since the suspension of the traditional face-to-face mode of schooling in March 2020, after the first case of the virus was recorded on the island. For the academic year 2019–2020, students did not attend school as usual for over three months (13 weeks), amounting to in-excess-of an entire school term, representing nearly 40% of an entire academic year (CAPRI, 2021).[7] After 19 months of physical closure, Jamaican children on aggregate lost an estimated 1.3 billion in-class hours.[8] To put this into context, according to the World Bank, a 5-month shut down of the global education system could result in an estimated loss of US $17 trillion in lifetime earnings based on a net present value calculations, or about 14% of today's global GDP.[9] The situation in Jamaica is 4 times worse as schools have been closed for almost 4 times as long given that the island is located in one of more severely affected regions. According to estimates emanating from the Ministry of Education, Youth and Information (MoEYI), about 120,000 children have been adversely affected by loss of learning.[10] These affected children account for a quarter (25%) of the total enrolment in public and primary schools on the island. UNESCO projects that over 100 million children globally will fall below the minimum proficiency level in reading due to the impact of school closures.[11]

The pandemic could also drive up learning poverty. The share of 10-year-olds who cannot read a basic text has increased from 53% before the

[7] CAPRI (September, 2021) Time Out The Impact of COVID-19 on Education https://www.capricaribbean.org/sites/default/files/public/documents/report/time_out_the_impact_of_covid_on_education.pdf.

[8] UNICEF. (2021). Reopen schools for all https://www.unicef.org/jamaica/reopen-schools-all.

[9] World Bank (December, 2021) Learning Losses from COVID-19 Could Cost this Generation of Students Close to 17 Trillion in Lifetime Earnings https://www.worldbank.org/en/news/press-release/2021/12/06/learning-losses-from-covid-19-could-cost-this-generation-of-students-close-to-17-trillion-in-lifetime-earnings.

[10] UNICEF. (2021). COVID-19 mash up everything https://blogs.unicef.org/jamaica/covid-19-mash-up-everything-my-life-without-school/.

[11] UNESCO. (March, 2021). One Year into COVID-19 education disruption: Where do we stand? https://en.unesco.org/news/one-year-covid-19-education-disruption-where-do-we-stand.

onset of the pandemic to around 70% in low- and middle-income countries during the pandemic.[12] Partial effects of the reduction in proficiency levels may have started to manifest in the Jamaican education system as evidenced by the average scores of the Primary Exit Profile (PEP) ability assessments. PEP has recently replaced the Grade Six Achievement Test (GSAT) as a more comprehensive appraisal of the readiness of students at the primary school level to transition to secondary education. PEP represents a shift from a regurgitated learning process to an assessment of analytical and critical thinking skills which has highlighted major shortcomings in the learning of students at the primary level in Jamaican schools. Based on the results of PEP, less than half of the students who sit in these exams are competent in Science (49%) and Mathematics (41%).[13] In contrast, a little over a half (55%) have achieved some degree of competency in Language Arts. By drilling further down into Language Arts, it showed that approximately 33% of students at the end of primary school could not read, 56% could not write and 57% could not identify information in a simple text (JETC, 2021). The latter gives insight into the level of learning poverty that exists on the island which is in line with global estimates for low- and middle-income economies but is at risk of being exacerbated by the COVID-19 pandemic.

Data suggests that the average assessment score deteriorated by 8% from 60% in 2020 to 52% in 2021 sitting CAPRI (2021). Global estimates suggests that the COVID-19 pandemic could result in a loss of between 0.3 (optimistic) and 0.9 (pessimistic) or an average of 0.6 year of schooling adjusted for quality, bringing down the effective average years of basic schooling that children receive during their schooling life from 7.9 years to 7.3 years (World Bank, 2020). In the Latin America and the Caribbean region, COVID-19 may reduce the baseline of 7.7 years to between 7.4 years (optimistic) and 6.8 years (pessimistic) (World Bank, 2020).[14] Jamaica's average years of schooling could also be reduced from

[12] World Bank. (October, 2021). World Bank: Pandemic Threatens to Drive Unprecedented Number of Children into Learning Poverty https://www.worldbank.org/en/news/press-release/2021/10/29/world-bank-pandemic-threatens-to-drive-unprecedented-number-of-children-into-learning-poverty.

[13] Jamaica Education Transformation Commission (September, 2021) The Reform of Education in Jamaica, 2021 Report https://opm.gov.jm/wp-content/uploads/flipbook/jetc-reform-of-education-in-jamaica-2021/.

[14] World Bank Group (June 2020) Simulating the Potential Impacts of Covid-19 School Closures on Schooling and Learning Outcomes—A Set of Global Estimates

9.7 years to anywhere between 9.4 to 8.8 years. Males are inclined to be more adversely affected as they already have been leaving school earlier than their female counterparts.

Globally, close to 7 million students including those at the primary and secondary level may be booted from the education system due to the income shock of the pandemic alone (World Bank, 2020). The world could lose as much as 16% of the investments governments make in this cohort of students' basic education throughout the pandemic. Similarly, the average student from the cohort in school today is projected to face an average reduction of between 2 and 8% of expected earnings every year. While the effect of the pandemic on education in Jamaica is not conclusive, these facts present a case of the deleterious impact it is likely to have, ultimately, creating a situation where only a fraction of the intended skills and lessons are learnt and a loss as it relates to cognitive development. As such, many may not be able to function properly when they become part of the employed labour force, lowering their human capital potential contributing less than they could to output and ultimately exacerbating Jamaica's arduous task of meeting its long-term developmental goals of achieving the status of a developed country by 2030 and having Jamaicans empowered to achieve their fullest potential.

The irregular school contact, learning loss and loss in years of schooling are sufficient indicators of the devastative impact of the COVID-19 pandemic on human capital development in Jamaica and many other small countries, especially, for those students who may never return to the education system after the pandemic. Where will they end up? How will they be absorbed into the labour force or the economy as adults? Their issue must be brought to the forefront. Already as shown in Fig. 5.4, 55.1% of Jamaicans who are outside of the labour force are there because they do not want to work. Many people want to receive pay without work, and they want to get rich quickly without education or the process of learning. Many of them without qualifications will end up in lower-level jobs working for minimum wage. Alessandrini and Milla (2021) found evidence to suggest that minimum wage impact human capital through years of schooling. Their results from research in Canada shows higher

https://openknowledge.worldbank.org/bitstream/handle/10986/33945/Simulating-the-Potential-Impacts-of-COVID-19-School-Closures-on-Schooling-and-Learning-Outcomes-A-Set-of-Global-Estimates.pdf?sequence=1&isAllowed=y.

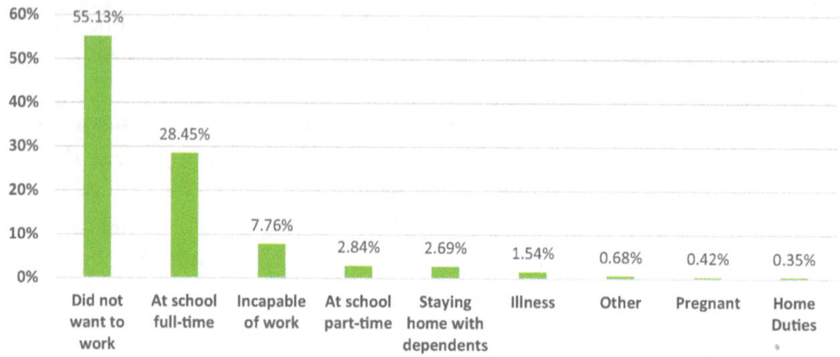

Fig. 5.4 People outside the labour force (*Source* STATIN)

minimum wage encourages enrolment in colleges for job-specific skills but not university.

5.2 Compensation, Real Wage and Productivity

5.2.1 Minimum Wage

Compensation of employees is very relevant to the level of productivity in Jamaica. Wages motivate people to work. In any country, the minimum wage is always a good place to start when trying to understand how its labour force is compensated. The minimum wage is a lower bound designed to encourage those with the least skills, education and training to participate in the labour market. It is a ration to reward those with the least ability to do work (Dobija, 2011). Minimum wage in Jamaica is $1.13 USD ($175 JMD) per hour 1/5 of the $7.25 USD per hour in the United States. Considering that between 60 and 65% of Jamaicans have no skill or education, it places them right at the minimum wage bracket. Of the 1.3 million people employed on the island, only 22% or 288,800 are professional, senior officials and technicians, leaving close to 1 million workers out of the high-income positions. Of this 78%, 300,000 are service workers, shops and market workers and 175,000 are elementary occupations, who more or likely earn close to minimum wage. These employees work to support themselves, their friends and their families who are either actively seeking work, retired, in school or unemployed. Based on the profession or the industry in which they work, some are better able to do so than others. Ivey (2017) estimated the impact of minimum wage increases on quality of life on 107 countries worldwide. Ivey (2017) found that increases in minimum wage marginally improve quality of life, ceteris paribus. As a result, many people although are employed, they earn less than what is required to live a decent life or to save for pension. Many must participate in the informal economy, corruption or receive remittances to compensate their incomes.

Perhaps if people were earning more, they could save more and have private pensions to support their retirement. Of the total employed in Jamaica, 65,500 are over the age of 65 (the retirement age) and should be collecting pension but are still working to cover their expenses. Jamaica has an ageing population which means that older people may have to work longer than they have anticipated. As people retire, there will be added pressures on pension funds for those who will be retiring as there are less people entering the labour force. This means that policymakers may have to investigate increasing the retirement age. Singapore, for example, extended their retirement age in 2021 to allow older people to work longer even though they have a thorough pension system. This opens the discussion of the value of real wages in the economy.

5.2.2 Real Wage and Productivity

> *Various reasons have been proposed to explain how wages may affect productivity. A sociological explanation is that lower-paid workers are less loyal to the firm. An explanation based on adverse selection is that a lower wage reduces the average quality of the work force because only the best workers quit. The most popular explanation of efficiency wages is "shirking". Because firms monitor effort imperfectly, workers sometimes shirk their responsibilities and risk getting fired; a lower wage reduced the cost of getting fired and thus raises the amount of shirking. In all of these efficiency wage theories, the impact of wages on productivity diminishes the incentive for a firm to cut wages in response to an excess supply of labor. If this productivity effect is sufficiently large, the normal competitive forces moving the labor market to the equilibrium of supply and demand are absent.* Mankiw (1990)

Jamaica has a population of 2.961 million people. Of which, less than half 1.327500 million people or 45% are employed to various industries dichotomized into two groups; those whose productivity is less than a $1million and those whose productivity is greater than $2.6 million, nothing in between. Of the 12 labelled industries, only three, Financial Intermediaries $2.6 million, Mining and Quarrying $2.8 million and Electricity, Gas and Water $3 Million, have productivity levels greater than 2.6 million, see Table 5.1. Correspondingly real wages in these industries are higher than the others although not homogenous in size relative to their levels of productivity. Real wages in both the Finance and Insurance and Mining and Quarrying are $2.8 million each on average per annum, while that of Electricity, Gas and Water is 1.9 million in 2019. Table 5.1 also shows that real wage to productivity ratios. Real wages are higher than productivity in Mining and Quarrying and Financial Intermediaries with ratios of 1.10 and 1.22 which are greater than 1, respectively. Electricity, Gas and Water pays less than their level of productivity. With a ratio of 0.65. Producers of Government Services are a standout; they are one of the least productive sectors but has the highest real wage ratio. Real wages are 4 times greater than the level of productivity. In other words, those working in government sectors' real wages are 4 times as high as their productivity. Alarming.

Construction, Wholesale and Retail, Manufacturing and Hotel and Restaurant also have ratios higher than one, but wages are below average income GDP per capita of $4800 USD.

In 2016, the government of Jamaica increased the income tax threshold to $1.5 million ($10,000 USD), way above the real wages of all the industries except the top 3. Between 2015 and 2019, 75% of tax paying Jamaicans excluding people who are self-employed earned JMD $1.5 million or less, placing them below the income tax threshold. By not paying any income taxes, employees in these industries have a higher disposable income from which to consume or save, see Fig. 5.5. Twenty-two percent (22%) earns between $1.5 to $6 million and only two (3%) earn above $6 million. This distribution of earnings complements the high level of untrained individuals that form the labour force in Jamaica. Of the employed, 23% are shop, market and service sale clerks in wholesale and retail with no training or qualification. Another 14% are in agriculture and 13% in elementary occupations also with little educational achievement. Recall that roughly 60% of the labour demand are basic qualification.

Incomplete labour markets lead to inefficiencies in labour allocation in Jamaica maintaining low wages. Low wages are likely to be a primary contributor to high levels of participation in the informal economy. Ivey (2017) found evidence to suggest that minimum wage earners who

Table 5.1 Real wage and productivity

Industry	*Real wage*	*Productivity*	*RW/P ratio*	*P\RW ratio*	*Variance*
Financial intermediation	$2,861,481.78	$2,605,179.13	1.10	0.91	$256,303
Mining and quarrying	$2,844,671.36	$2,340,097.75	1.22	0.82	$504,574
Electricity, gas and water Supply	$1,975,884.73	$3,019,075.10	0.65	1.53	−$1,043,190
Producers of government services	$1,752,776.95	$433,349.65	4.04	0.25	$1,319,427
Construction	$866,451.07	$510,855.34	1.70	0.59	$355,596
Manufacturing	$757,327.13	$666,865.36	1.14	0.88	$90,462
Real estate renting and business activities	$752,863.72	$890,814.33	0.85	1.18	−$137,951
Transport storage and communication	$744,322.93	$930,573.04	0.80	1.25	−$186,250
Wholesale and retail repair of motor vehicle and equipment	$608,815.62	$501,745.73	1.21	0.82	$107,070
Hotels and restaurants services	$452,012.62	$431,970.22	1.05	0.96	$20,042
Agriculture, hunting, forestry and fishing	$178,064.17	$246,663.94	0.72	1.39	−$68,600

Fig. 5.5 Salary bands in Jamaica (*Source* Tax office authority of Jamaica)

are experiencing relatively high standards of living were either being supported by family members who are abroad in the form of remittances or by the participation in black market or illicit activities. ILO (2018) confirms that the levels of education and training are key factors affecting the level of informality. ILO (2018) revealed that there is an inverse relationship between education and informality. When the level of education increases the level of informality decreases. Additionally, people who have completed secondary and tertiary education are less likely to be in informal employment compared to workers who have either no education or completed primary education.

5.2.3 Assessing the Relationship Between Compensation and Productivity in Jamaica

> *A pure slowism a rule di country now you know" No body a work fi nothing now. Di more yu work is di less yu get, but everyday yu a go a di shop everything raise. If we not getting justice we just have to slow up things. No production. Dat is two way justice.*[15]

5.2.3.1 Background

There has been a periodic if not consistent debate in Jamaica relating to the effort that is required for the satisfactory completion of jobs relative to the compensation that is received for the effort exerted. This debate has been going on for decades. Yet, the evidence suggests that there have not been any concrete solutions that would resolve the issue and simultaneously foster sustained productivity growth for Jamaica and decent living standards for its people. As a result, different factions of the labour force continue to engage in industrial actions which are chiefly dominated by strikes. Furthermore, these workers tend to operate in some of the most critical areas of the economy which can potentially cripple the already depressed and ever eroding productivity levels of the nation. A common sentiment is that successive governments tend to find resources for other activities but not to help balance the quality of life of Jamaicans, especially in a COVID-19 depressed environment coupled with the spillover effects

[15] Carter. (1997). Why Workers Won't Work: The Worker in a Developing Economy: a Case Study of Jamaica https://books.google.com.jm/books?id=6hNGAAAAYAAJ&q=why+workers+wont+work&dq=why+workers+wont+work&hl=en&sa=X&redir_esc=y.

from the Russia-Ukraine war and supply chain disruptions emanating from China.[16] These events are responsible for increased inflationary pressures which have started to erode the purchasing power and standard of living of many people globally. Jamaicans are not exempted. While Jamaica has not reached its tipping point, countries like Sri Lanka have succumbed to economic pressures which have resulted in the Prime Minister's resignation being tendered. Jamaica can learn from the experience of these countries. Ultimately, Jamaicans want to see their pay reflect their self-worth and the value of their labour and this can be achieved with the design and implementation of a broad-based productivity linked compensation system which will both drive productivity levels and eventually better lives for Jamaicans.

In order to determine if a productivity linked compensation system would be beneficial to Jamaica, we must first determine if there is a significant connection between compensation and productivity. The literature on the link between compensation and productivity displays varying results where some researchers, for example Lawrence (2016) shwas that productivity has been growing over time, but wages have stagnated. Others for example, Strain (2019) posits that the link is very much intact.[17] Forfack and Temkeng (2021)[18] analysed the movement of productivity and compensation in 25 countries across Europe which reveals that productivity grows at a faster pace relative to compensation, thereby creating a gap. While a gap exists, Forfack and Temkeng (2021) found evidence to suggest that the variables are linked in the long run and revert to their long-run Eq. (5.1) equilibrium following short-run shocks. Stansbury and Summers (2018)[19] investigated the link between compensation and productivity in the United States for the period 1973 to 2016. Their results confirmed that compensation and productivity are

[16] Different administration have spent millions on a state dinner which is viewed as a slap in the face for workers who have not seen inflation eroded their real wage.

[17] Strain, M. R. (2019). The Link between wages and productivity is strong. *Expanding Economic Opportunity for More Americans, The Aspen Institute*. https://www.aei.org/wp-content/uploads/2019/02/The-Link-Between-Wages-and-Productivity-is-Strong.pdf.

[18] Fofack, A. D., & Temkeng, S. D. (2021). A cross-sectoral analysis of the relation between labor productivity and labor compensation in the European Union. *Applied Econometrics*, *62*, 54–65.

[19] Stansbury, A. M., & Summers, L. H. (2018). *Productivity and Pay: Is the link broken?* (No. w24165). National Bureau of Economic Research. https://www.nber.org/system/files/working_papers/w24165/w24165.pdf.

linked for the United States. In more granular details, Stansbury and Summers (2018) results show that a one percent improvement in productivity growth is related to a 0.7 to 1 percentage point higher median and mean compensation growth, respectively. They also show that a one percent increase in productivity is associated with a 0.4 to 0.7 percentage point higher compensation for production/non-supervisory workers.

5.2.3.2 *Data and Methodology*

To get an understanding of the compensation-productivity relationship in Jamaica, we follow the empirical methodological approach estimated by Stansbury and Summers (2018) as shown by Eq. (5.1) in its simplest form.

$$compensation\ growth_t = \alpha + \beta\ productivity\ growth_t \tag{5.1}$$

In this setting, $\beta = 1$ represents the perspective of the complete "linkage" between compensation and productivity while $\beta = 0$ represents complete "delinkage" where the continuum of β ranges from 0 to 1.

We use a sample of annual data over the 1993–2019 period. Similar to Standbury and Summers (2018), we compute the three-year moving averages for real compensation,[20] labour productivity and unemployment for our baseline model. We then estimate the econometric model specified in Eq. (5.2) where the three-year moving average of the change in log of real compensation is regressed on the three-year moving average of the change in log labour productivity as well as the current and lagged values of the three-year moving average of the unemployment rate. Similar to Standbury and Summers (2018), we included the unemployment rate to capture the fluctuations in the business cycle which can give an indication of the state of the economic conditions in Jamaica.

$$\begin{aligned}\frac{1}{3}\sum_{t}^{t+2} \Delta logComp_{t-i} =& \alpha + \beta\frac{1}{3}\sum_{t}^{t+2} \Delta LogProd_{t-i} \\ &+ \gamma\frac{1}{3}\sum_{t}^{t+2} Unemp_{t-i} + \delta\frac{1}{3}\sum_{t}^{t+2} Unemp_{t-i-1} + \varepsilon_t\end{aligned} \tag{5.2}$$

[20] This is defined as total remuneration, in cash and in kind, received by employees from an employer for work undertaken during the accounting period. For more details see https://statinja.gov.jm/Methodologies/STATIN_National_Accounts_Methodology_2015_10.pdf.

5.2.3.3 *Results and Discussion*

Our results suggest that there is a strong link between real compensation[21] and labour productivity in Jamaica. To be more precise, it was estimated that a 1% increase in labour productivity growth increases real compensation by 0.9% (See Table 5.2). This is a stronger link compared to the 0.7 estimated by Stansbury and Summers (2018)[22] for the United States. These findings suggest that the same level of productivity in Jamaica relative to the United States should result in higher compensation in Jamaica. Stansbury and Summers (2018) also indicated that while productivity growth seem to have been labouring to elevate compensation levels in the United States, other factors are likely to have been acting as suppressants. Interestingly, this is similar to what Forfack and Temkeng (2021) found in Europe but more specifically to countries within the EU. An interesting discovery that should not go unnoticed is that while the link between compensation and productivity are found to be reasonably strong in these countries, productivity growth in Jamaica relative to the United States and some countries with the EU is weaker. In fact, the evidence shows that Jamaica's productivity has been declining which is likely to be a major factor in the erosion of the real wage per person (purchasing power) of Jamaicans overtime as shown in Fig. 5.6 (Fig. 5.7).

Between 1994 and 2007 before the Global Financial Crisis GCF hit Jamaica, Real Compensation, Labour Productivity and Real Wage grew at average rates of 2.5, 0.04 and 1.42%, respectively, all of which subsided during the crisis; Real Compensation growth fell to −2.49, Labour Productivity fell to −1.04 and Real Wage fell to −5.63. After the crisis, these growth rates improved but remained negative with no signs of arrive at pre crisis levels (Table 5.3).

5.2.3.4 *Post Diagnostic Tests*

It is critical that the Classical Linear Regression Assumptions are all satisfied. To ensure robustness of the estimates, a series of post estimation diagnostic tests are explored. Based on our assessment, we can conclude that our model is robust as evidenced by the Ramset Retest (0.8453) for omitted variable bias, Jarque-Bera (0.8174) for heteroskedasticity, Breush

[21] Compensation was deflated using the Consumer Price Index in order to account for changes in the livings standards of Jamaicans.

[22] Stansbury, A. M., & Summers, L. H. (2018). *Productivity and Pay: Is the link broken?* (No. w24165). National Bureau of Economic Research.

Table 5.2 Compensation and productivity regression output

Variables	*RM1*	*RM2*	*RM3*
Log Productivity, 3 year moving average (Change)	0.977* (0.513)	0.852* (0.482)	0.900*** (0.277)
Unemployment, 3 year moving avg		0.0248 (0.0207)	0.0641*** (0.0181)
Lagged unemployment, 3 year moving avg			−0.295*** (0.0371)
Constant	0.0145*** (0.0043)	−0.0499 (0.054)	−0.157*** (0.0481)
Observations	24	24	24
R-squared	**0.164**	**0.193**	**0.787**

Note Dependent variable is the 3-year moving average of log compensation. Robust standard errors in parentheses ***$p < 0.01$, **$p < 0.05$, *$p < 0.1$

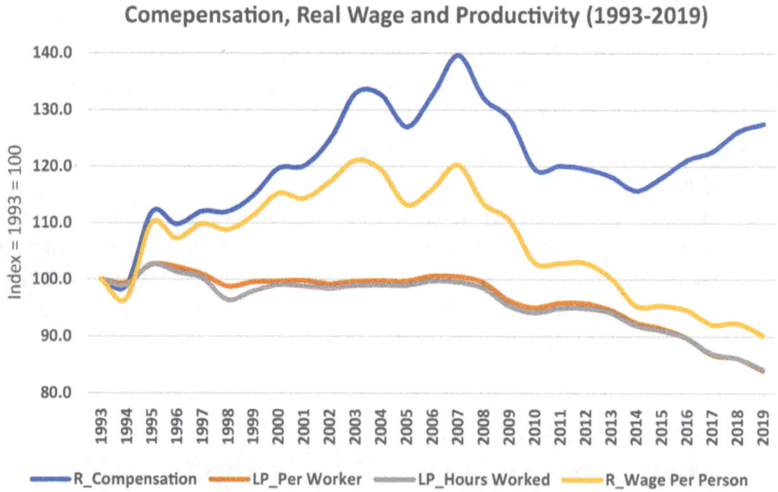

Fig. 5.6 Compensation, Real Wage and LP (1993–2019) (*Source* STATIN)

Pagan (0.4393) and Shapiro Wilk tests normality of the fitted values of the regression residuals, and lastly the Variance Inflation Factor (VIF) which has an average variance of 1.2 which suggest no multicollinearity.

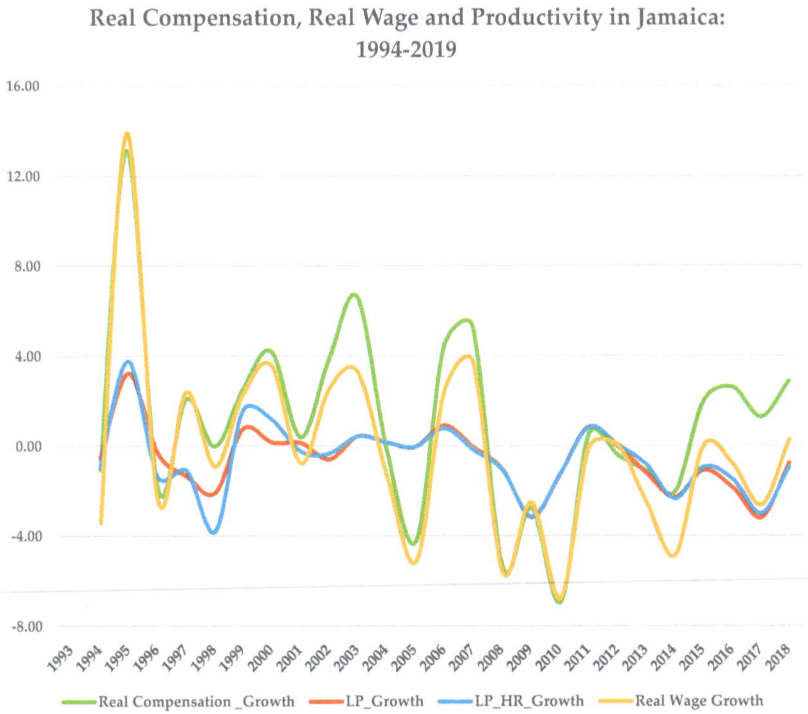

Fig. 5.7 Real compensation, Real Wage and Productivity (Growth) in Jamaica (*Source* STATIN)

Table 5.3 Real Compensation, Real Wage and Productivity Growth Rates in Jamaica (1993–2019)

Years	*Comp*	*LP*	*LPHW*	*RW*
1994–2007 (Pre-crisis)	2.49	0.04	−0.02	1.42
2008 (Crisis)	−5.42	−1.04	−1.06	−5.63
2009–2019 (Post-crisis)	−0.28	−1.52	−1.41	−2.04

5.2.3.5 *Conclusion and Recommendation*

In this chapter, we take a look first at Jamaica's human capital levels, then the link between human capital and productivity. Finally, we analyse the link between productivity and compensation. Jamaica has low levels of human capital that has limited the productivity of the country. To analyse the relationship between productivity and compensation, we employ a model proposed by Stansbury and Summers (2018). The results suggest that compensation and productivity are strongly linked in Jamaica. Even though compensation increases sometimes when productivity is falling, to achieve a sustainable increase in compensation-productivity must increase. To achieve this increase in productivity, other factors such as human capital plays an integral role. Therefore, it is prudent for the government of Jamaica to make the necessary effort to remedy the low human capital levels in the country.

References

Alessandrini, D., & Milla, J. (2021). Minimum wage effects on human capital accumulation: Evidence from Canadian Data.

Australian Goverment. (2017). *Productivity review*. Discussion paper. https://www.pc.gov.au/inquiries/completed/productivity-review#report

Becker, G. S. (1993). Nobel lecture: The economic way of looking at behavior. *Journal of Political Economy, 101*(3), 385–409.

CAPRI. (2021, September). *Time out the impact of COVID-19 on education*. https://www.capricaribbean.org/sites/default/files/public/documents/report/time_out_the_impact_of_covid_on_education.pdf

Dobija, M. (2011). Labor productivity vs. minimum wage level. *Modern Economy*, *2*(5), 780–787.

Fofack, A. D., & Temkeng, S. D. (2021). A cross-sectoral analysis of the relation between labor productivity and labor compensation in the European Union. *Applied Econometrics, 62*, 54–65.

Fogel, R. (1999). Catching up with the economy. *American Economic Review, 89*(1), 1–21.

Garavan et al. (2001). *(19) (PDF) Human capital development and its impact on firm performance: Evidence from developmental economics*. https://www.researchgate.net/publication/26628217_Human_Capital_Development_and_Its_Impact_on_Firm_Performance_Evidence_from_Developmental_Economics [accessed Nov 15, 2021].

Global Competitiveness Report Special edition 2020.

International Labour Organisation (ILO) Report. (2018). *Women and men in the informal economy: A statistical picture* (3rd ed.). ISBN: 9789221315810. https://www.ilo.org/global/publications/books/WCMS_626831/lang--en/index.htm

Ivey, W. (2017). An empirical analysis of minimum wage on the quality of life. *Southwestern Journal of Economics, 12*(1).

Jamaica Education Transformation Commission. (2021, September). *The reform of education in Jamaica, 2021 report*. https://opm.gov.jm/wp-content/uploads/flipbook/jetc-reform-of-education-in-jamaica-2021/

Lawrence, R. Z. (2016). Does productivity still determine worker compensation? Domestic and international evidence. In *The US labor market-questions and challenges for public policy* (pp. 42–62).

Maneejuk, P., & Yamaka, W. (2021). The impact of higher education on economic growth in ASEAN-5 countries. *Sustainability, 13*(2), 520.

Marimuthu, M., Arokiasamy, L., & Ismail, M. (2009). *Human capital development and its impact on firm performance: Evidence from developmental economics*.

Myrdal, G. (1971). The economic impact of colonialism. *Developing the Underdeveloped Countries*, 52–57.

Stansbury, A. M., & Summers, L. H. (2018). Productivity and Pay: Is the link broken? (No. w24165). National Bureau of Economic Research. https://www.nber.org/system/files/working_papers/w24165/w24165.pdf

Strain, M. R. (2019). The link between wages and productivity is strong. Expanding economic opportunity for more Americans, The Aspen Institute. https://www.aei.org/wp-content/uploads/2019/02/The-Link-Between-Wages-and-Productivity-is-Strong.pdf

Sweetman, A. (2002). Working smarter: education and productivity. *The review of economic performance and social progress*, *2*. ASEAN's economic growth|ASEAN NOW Report (austrade.gov.au) https://www.ilo.org/global/about-the-ilo/newsroom/news/WCMS_627189/lang--en/index.htm#:~:text=The%20report%20shows%20that%2093,over%20740%20million%20are%20women.

CHAPTER 6

Migration and Human Capital Flight

6.1 Migration Patterns from Latin America and the Caribbean

Globalization has increased the interdependence of countries by facilitating greater cross-border movement of business, investment, goods and services, financial assets, technology and people. People tend to migrate from developing countries to larger economies such as those of the United States, Europe, the Russian Federation, the United Arab Emirates and Saudi Arabia (IOM, 2019). In the case of Latin America and Caribbean (LAC), people migrate mainly to North America, Europe and other developed countries to a lesser extent. The number of people from the LAC region that have migrated to North America has increased by 166% over the last 30 years, from an estimated 10 million in 1990 to 26.6 million in 2019 (IOM, 2019). It is also estimated that Europe holds approximately 5 million people from the LAC region which has slightly increased since 2015. The number of migrants from the LAC region overall has more than quadrupled since 1990 (IOM, 2019). With projected increases in the populations of many developing subregions and countries in the coming decades, the migration pattern is estimated to remain the same for many years in the future (IOM, 2019). Globally, it was estimated that approximately 272 million people migrated in 2018,

A. Haughton and W. Ivey, *Overcoming Productivity Challenges in Small Countries*, https://doi.org/10.1007/978-3-031-23301-2_6

nearly two-thirds being labour migrants.[1] Roughly, 3.5% of the world's population is increasing (IOM, 2019).

Migration has consequences for both home and host countries. Labour that migrates adds valuable human capital resources to the labour force of the host countries and contribute to the brain drain of their country of birth. Substantially, people who immigrate contribute to the economy of the host country through three channels: 1. they add to the labour market, 2. they increase the taxable population and 3. they also increase economic competitiveness. Migrants, especially if selected through a screening process, as is the case with Canada, for example, bring the labour skills and technical competence needed to fill shortages in the foreign labour market. They make labour markets more competitive and therefore increase the efficiency and productive output, while reducing the efficiency, competitiveness and productivity of the labour market in the countries they migrate from. As our analysis will show, some of the most educated people migrate leading to human capital flight from home countries. Many Jamaicans migrate to other countries in search of opportunities for better employment, higher wages, better working conditions and safety. The effect of which can be significant.

Many of the most skilled and professionally trained people emigrate to more advanced countries like the United States, Canada and UK where they are better compensated and feel a greater sense of job satisfaction. Internationally, they develop good work ethics encouraged by the productive environment offered by more developed countries. Domestically, they would otherwise struggle to survive or sometimes encounter envy and grudge from their less fortunate peers. They would ultimately migrate and contribute to the human capital accumulation of these advanced countries at the expense of the brain drain of small countries like Jamaica and many other SIDs. The process of reallocating productive labourers erodes the competitiveness of these small nations which can send them down a spiral of low productivity and weak economic growth. Low output prevents a decent standard of living for many employed while many of the unemployed are in poverty in Jamaica and other SIDS. These countries are caught in an underdevelopment trap. The main advantage of migration is that they send back remittances to their home countries, which to some extent act as a deterrent to working locally and do not contribute to the

[1] International Organization of Migration. (2019), World Migration Report 2020. https://publications.iom.int/system/files/pdf/wmr_2020.pdf.

productivity of the island. Jamaica and many other small countries are in this position of underemployment of labour and must figure out how to become more productive.

However, despite threats to globalization arising from major economic and geopolitical occurrences, for example, Brexit in Europe, America and China's economic repatriation policies and the restriction of movement arising from the COVID-19 pandemic, people will continue to relocate. "Pull" and "Push" factors motivate people to move from their country of birth to another country. Pull[2] factors are circumstances in the destination country that make it a more attractive place to live relative to their home countries.[3] On the other hand, push factors describe the conditions in a migrant's home country that makes it difficult or even impossible to live there. Many SIDS including Jamaica are riddled with push factors like high levels of crime and violence, low-income jobs, weak legal and justice systems, poor health care and education systems that force people to be desperate to leave. While the more developed countries have accommodating pull factors like better opportunities, higher incomes, better health care and education systems that entice foreigners to come and live, many developed countries even recruit for the specific type of migrant labour that their economy need. A country like Canada constantly advertises their permanent residency to skilled or educated people from developing countries.

Castelli (2018) explains that migration is a complex phenomenon, where "macro", "meso" and "micro" factors act together to inform the final individual decision to migrate. *Firstly*, political, demographic, socio-economic and environmental situations are described as "macro" factors which are the main drivers of forced migration, either internationally or locally. These are largely out of people's control. SIDS that are highly politically unstable, and/or those plagued with corruption like Jamaica, might force people to migrate. Demographic and socio-economic factors, like sex, age, income, wealth and educational levels, also contribute to

[2] Pull factors include opportunities for better employment, higher wages, better working conditions and attractive amenities as noted by Thet (2014).

[3] National Immigration Forum (July, 2019) Push or Pull Factors: What Drives Central American Migrants to the U.S.? https://immigrationforum.org/article/push-or-pull-factors-what-drives-central-american-migrants-to-the-u-s/#:~:text=Push%E2%80%9D%20factors%20are%20conditions%20in,factors%20include%20violence%2C%20gender%20inequality%2C.

people's decision to migrate. Across the developing world, the most basic economic factors which motivate migration are low productivity, unemployment and underdevelopment, poor economic conditions, lack of opportunities for advancement, and exhaustion of natural resources and natural calamities (Thet, 2014). *Secondly*, "meso-factors" include communication technology and diasporic links as attractive factors. Social media plays a crucial role at this level as it attracts people out of their origin countries by elevating awareness of living conditions in the affluent world. However, Castelli (2018) notes that the social media effect is often exaggerated. Nevertheless, people from the developing world acquire a favouritism for the lifestyle they see others enjoying on TV or social media and yearn to leave their circumstances and do the same. In Jamaica, getting an American visa is a major accomplishment for the average citizen let alone a Green Card or citizenship. Many of the people are foreign minded. But it is hard to blame them considering the circumstances and standard of living that they experience. Lastly, "micro-factors" include education, religion, marital status and personal attitude to migration. Castelli (2018) posits that there is an existing stereotype as it relates to the illiterate, poor and rural migrant reaching the borders of affluent countries which must be disbanded. Castelli (2018) notes that the poorest people simply are not endowed with the means to escape war and poverty and as such they remain trapped in their country or in the neighbouring ones. The aim of this chapter is to understand the reasons why people migrate from Jamaica as well as the reason why they might return to Jamaica in the future.

6.2 Migration Survey

6.2.1 Methodology and Data

To garner a better understanding of the reasons why people migrate and the reasons for which they might return, a survey instrument was designed and administered. In the context of the global health pandemic, social distancing was introduced as a containment strategy to slow the spread of the coronavirus, coupled with the unique ability of online based surveys to quickly reach potential respondents across the globe, and the researchers utilized Survey Monkey, an online data collection platform, to collect data from people in the diaspora who have migrated from Jamaica. The survey instrument was a questionnaire consisting of 15 open- and close-ended questions. The instrument was administered through a web link for a 1-month period, from 15 March to 15 April 2021 on social media platforms. In distributing this instrument, the non-probability snowball sampling technique was used. In essence, the survey instrument was sent to individuals who have migrated and now form part of the Jamaican diaspora. These individuals were then asked to share the survey instrument link with people within their network who would have also migrated from Jamaica. The next section outlines and discusses the findings.

6.2.2 *Results*

The survey instrument had an average response rate of 96% and a maximum of 291 respondents.[4] The results of the survey indicated that 63.2% of the respondents were females while males accounted for 36.8%. The average age of the respondents is approximately 32 years old which ranges from 15 to 70 years old. The result of this survey also shows that most of the respondents resided in the United States (58.3%) and Canada (14.7%), followed by China (9.1%) and the UK (8.8%). This corresponds with the historical migration data from the PIOJ which shows that the primary destinations of Jamaican emigrants are the United States, UK and Canada. Respondents in this survey would have migrated as early as 1971 and as late as 2021, a range of 5 decades. On average, a respondent from this survey would have migrated in 2013. As it relates to educational attainment, approximately 81.1% of the respondents possessed tertiary-type education. Undergraduate qualifications accounted for 39.2%, master's degree (30.9%), doctoral (2.75%) and short tertiary-type certificates or equivalent accounted for 8.3%.

Figure 6.1 shows the dispersion of where these Jamaicans were currently living at the time of data collection. Majority of the respondents were residing in the United States which is consistent with surveys conducted by other intuitions. This is not surprising since most Jamaicans do migrate to the United States. While some Jamaicans intend to return to live out their older days, others are more reluctant.

[4] The typical time spent to complete the survey was three (3) minutes on average.

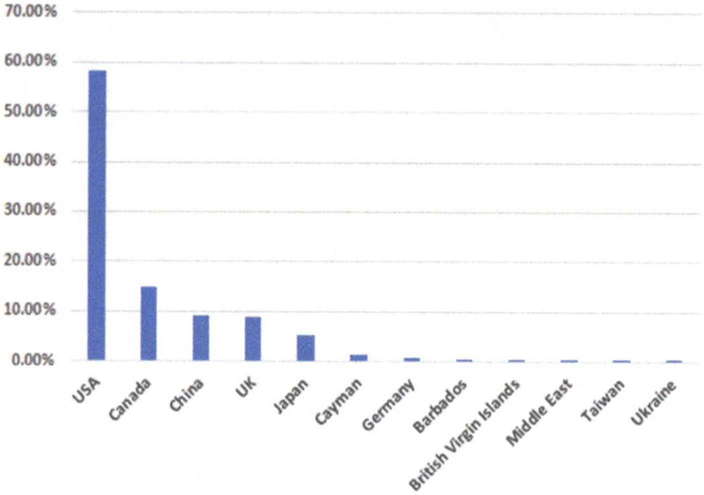

Fig. 6.1 Where respondents are currently living

6.2.3 The Pull Factors

Pull factors were instrumental reasons why these individuals opted to migrate to a foreign country. They were more macros including socio-economic and political reasons as well as micro, personal and family reasons. The survey results indicate that roughly 35.5% of respondents left Jamaica to find better employment opportunities abroad. About 27.5% migrated to settle for other reasons, 25%.4% left to receive better education or training and 7.6% of respondents migrated for marriage. The desire to migrate for better employment opportunities seems to be a plausible reason for the respondents to seek alternative ways to sustain their livelihoods as approximately 51% of respondents declared that they did not have a full-time job while they were in Jamaica while 49% noted that they had full-time employment.

Since people migrated at different times, it is also imperative to understand what they were engaged in in terms of earnings and the type of human capital that the country is losing. The results suggest that 43% of the respondents did not have a job at departure while the remaining 57% of occupations at departure are shown in Table 6.5. Teachers, finance professionals, students, accountants and nurses are the top 6 professions captured by the survey. While these professions accounted for approximately 46% of respondents who have migrated, the occupation category is widely dispersed and captures up to 60 different occupations. In essence, people are leaving Jamaica regardless of their current occupation to find better job opportunities. Many of these migrants more than quadruple their income after migrating.

The survey results indicate that the average monthly income earned by respondent while working in Jamaica was estimated at $107,457.50 (800 USD) relative to $703,886.20 ($5,000 USD) when they migrated, see Table 6.1. The salary earned in Jamaica relative to a foreign country is 6.6 times lower on average which explains why foreign countries can attract the highly trained Jamaicans who have migrated and those who will continue to (Fig. 6.2).

Table 6.1 Monthly income

Variable	*Average*	*Std. Deviation*	*Minimum*	*Maximum*
Jamaica_Income	$107,457.50	$78,190.23	$1000	$500,000
Foriegn_Income	$703,886.20	$608,410	$704,79.51	$4,950,181

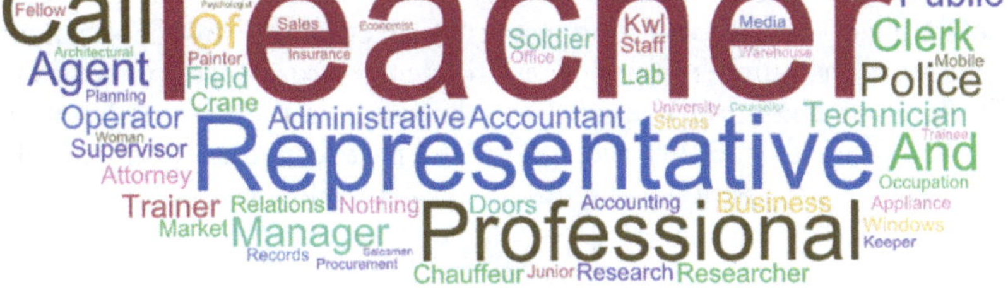

Fig. 6.2 Occupation type whilst abroad

Fig. 6.3 Push factors

6.2.4 *The Push Factors*

The Push factors were also macro and micro, less meso factors were identified. People migrate and don't return for a variety of reasons included in the Push factors; Crime, Violence, and Quality of Life. According to the survey respondents, the state of the economy, income, corruption, safety and poverty also push people to migrate. Cost of living, jobs, better opportunities abroad, "Bad-mind" a colloquial term for envy & grudge is also a reason why Jamaicans opt to migrate (Fig. 6.3).

The United States (US), UK (UK) and Canada are the main beneficiaries of the bulk of Jamaican migrant labour although they are not the sole beneficiaries as over the last few years. Jamaicans have started to take advantage of useful opportunities in Asian countries such as China and Japan. Migration from Jamaicans is nothing new. Jamaicans have been migrating to the United States since the nineteenth century (Chung et al., 2009). Between 1971 and 2004, 571,265 Jamaicans migrated to the US. Between 2002 and 2018, twenty-five thousand (25,000) Jamaicans on average have migrated to the US, UK, and Canada annually, see Fig. 6.4.[5] Of the 25,000 who migrate on average per annum roughly 75% goes to the US, 15% to Canada and 10% to the UK. However, in recent times Canada has become more attractive as evidenced by the sustained increases in the number of permanent residency cards issued to Jamaicans. The level of migration is expected to increase as the current level has trended above the 25,000 average for the past few years. Cumulatively, over the 17-year period roughly 417,327 people, more than 10% of

[5] Cumulatively, it is estimated that between 2002 and 2018, migration to the US, UK and Canada have amounted to 327,431, 46a,090, and 43,806, respectively.

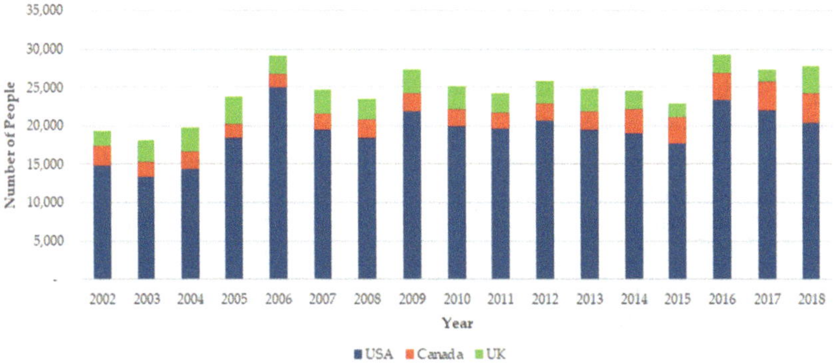

Fig. 6.4 Jamaica's Emigration to Canada, UK and US (*Source* Planning Institute of Jamaica)

Jamaica's population, have migrated. For more context, 27% of Jamaica's combined population is migrants which places Jamaica among the top 20 countries in the world with the highest proportion of its combined population being emigrants (IOM, 2019).

6.3 Human Capital Flight

Human capital is considered as one of the fundamental drivers of economic development and sustainable competitive advantage in the global marketplace (Younas and Ahmad, 2020). Human capital flight or brain drain can be described as a situation where the smartest and best educated people in a particular society or country leave to settle in another society or country.[6] Haque and Kim (1995) developed an endogenous growth model with heterogeneous agents which showed that "human capital flight" or "brain drain" can lead to a permanent reduction in income and growth of the country of emigration relative to the country of immigration. Classical theoretical studies on Human Capital Flight also hold that the emigration of highly educated people is beneficial

[6] Brain Drain Definition https://financial-dictionary.thefreedictionary.com/Human+capital+flight.

for destination countries and harmful for source ones.[7] Liao et al. (2017) posits that this type of education-based migration is a larger contributor to the enhancement of per capita output relative to work-based migration. Once a significant level of brain drain is initiated, it can have damaging and lasting effects on the economy that induce further waves of emigration by high-skilled workers (Docquier, 2014). Typical examples include Iran after the 1978–1979 revolution, the Irish crisis of the early 1980s, the ex-Union of Soviet Socialist Republic (USSR) after 1991 and in more contemporary times Puerto Rico following their debt Crisis which started in 2014. Since the 1990s, brain drain rates have charted an upward trajectory and the most severe increases have been noticed in countries that are small, poor and tropical.[8] Haiti and Jamaica are two good Caribbean examples. They are two of the most severely affected countries in the world as it relates to the net emigration of college graduates, excluding small states with a workforce less than one million workers where the population age exceeds 25 years old. These countries experience net emigration rates of more than 80% where the brightest minds go to more advanced economies (see Table 6.2). The other 8 countries within the top 10 are estimated to be losing between 41 and 53% of their college graduates. These countries include Liberia, Sierra Leone, and Somalia (Africa) or Laos and Afghanistan (Asia) among others. With high leakages of human capital, the capacity to and grow and develop sustainably may be impaired.

Women are natural caregivers who help to shape the minds and instil positive values, attitudes and morals into the fibre of their children. A path towards prosperity. Hence, their role in the development of a nation is crucial and should not be underestimated. This renders the emigration of highly skilled women a very concerning matter. It is recognized that women's human capital is an important determinant of labour productivity, children's education and economic growth (Artuc et al., 2014). Countries that experience negative net emigration rates of high-skilled female labour are more likely to grow at a slower pace and see lower increases in income levels relative to other countries that experience positive net emigration rates. The situation can be further exacerbated if

[7] Corrado Di Maria Migration & Piotr Stryszowski (2006), Human Capital Accumulation and Economic Development http://citeseerx.ist.psu.edu/viewdoc/download?doi=10.1.1.514.3533&rep=rep1&type=pdf.

[8] https://wol.iza.org/articles/brain-drain-from-developing-countries/long.

Table 6.2 Total brain drain: most affected countries in the world

Country[9]	*Net (%)*	*Gross (%)*	*Non-OECD (%)*
Jamaica	84.6	85.6	0.6
Haiti	80.9	81	5.9
Liberia	53.3	59.8	13
Sierra Leone	51.9	53.4	12.7
Eritrea	49.2	49.9	44.6
Laos	45.5	45.8	0.5
Somalia	45.1	46.2	26
Afghanistan	44.1	45	45.5
Bosnia and Herzegovina	41.7	44.6	51.5

governments in small or developing nations make significant investments in the training and educating of their female labour force but are not able to retain a significant proportion of the trained labour. Arctic et al. (2014) estimates the brain drain of women across the globe. It turns out that Jamaica topped the ranking for the most severely affected country in the world in the context of women brain drain (Table 6.3).[10] The island has a net emigration rate of 87.7%. In other words, less than 1/5th of the output of female college graduates stays in Jamaica. The other countries rounding out the top 10 are predominantly from Sub-Saharan Africa.

The percentage of tertiary educated people that emigrate from the Caribbean region ranks among the highest in the world (Wenner, 2016). Approximately 70% of Caribbean people with tertiary education have migrated to advanced countries.[11] Docquier and Schiff (2008) estimated that small states have a brain drain rate five times that of all developing countries, twelve times that of high-income countries and eight times the world average (Table 6.4).[12]

[9] Artuc, E., Docquier, F., Özden, Ç., & Parsons, C. (2014). *A global assessment of human capital mobility: the role of non-OECD destinations*. The World Bank. https://www.econstor.eu/bitstream/10419/107527/1/dp8746.pdf.

[10] The non-OECD share measures the share of non-OECD countries in gross emigration of college graduates.

[11] Mark Wenner (September, 2016) Brain Drain: A Curse of Small States? https://blogs.iadb.org/caribbean-dev-trends/en/brain-drain-a-curse-of-small-states/.

[12] Frederic Docquier and Maurice Schiff (March, 2008). Measuring Skilled Emigration Rates: The Case of Small States. http://ftp.iza.org/dp3388.pdf.

Table 6.3 Women's brain drain: most affected countries in the world (2000)

Country[13]	*Net (%)*	*Gross (%)*	*Non-OECD (%)*
Jamaica	87.7	88.6	0.7
Haiti	83.3	83.5	3.4
Sierra Leone	73.7	75.9	9.6
Liberia	68.7	75.8	8.8
Afghanistan	54.3	54.8	34
Laos	51.4	51.7	0.5
Cameroon	51.1	61.3	17.3
Congo, Rep. of the	49.3	51.2	18.9
Kenya	48.6	55.1	8.5
Somalia	45.5	46.4	19.3

Table 6.4 Emigration rates by educational attainment and country of birth 2000

Country[14]	*Tertiary*	*Secondary*	*Primary*
Jamaica	82.5	30	8.3
Haiti	81.6	27.5	2.5
Trinidad and Tobago	78.4	20.6	6.1
Saint Kitts and Nevis	71.8	37.1	10.3
Antigua and Barbuda	70.9	35.9	6
Grenada	66.7	69.5	9.9
Barbados	61.4	24.3	9.9
Dominica	58.9	60.6	8
Saint Vincent and the Grenadines	56.8	53.4	6.3
Bahamas, The	36.4	12.1	1.5
Saint Lucia	36	32.1	2.6
Cuba	28.9	9.4	5.5
Dominican Republic	21.7	30.9	5.8
The Caribbean	40.9	17.8	5.1

[13] Artuc, E., Docquier, F., Özden, Ç., & Parsons, C. (2014). *A global assessment of human capital mobility: the role of non-OECD destinations*. The World Bank. https://www.econstor.eu/bitstream/10419/107527/1/dp8746.pdf.

[14] Docquier, F., & Marfouk, A. (2004). *Measuring the international mobility of skilled workers (1990–2000): release 1.0*. The World Bank. https://openknowledge.worldbank.org/bitstream/handle/10986/14126/wps3381.pdf?sequence=1.

Human capital flight has been a consist feature of Jamaica's structural make-up and may persist if qualified and skilled people find opportunities in foreign countries more attractive that what exists in Jamaica. Highly skilled workers and professionals have been leaving the island since the 1960s. Many commentators in this era attributed the emigration to high levels of taxation and economic hardships which were rooted in the failure of democratic socialist policies. A more pellucid analysis of the migration trends in previous years revealed that not only that the high rates of emigration were part of an ongoing trend set soon after World War II, but also that the various annual fluctuation in emigration rates corresponded more to immigration policies of the receiving countries than to the policies of the sending nation (Cooper, 1985).[15] In 1989, output from training institutions in Jamaica amounted to 20,752 trained people. This was primarily underpinned by a 17.1% increase for the professional, technical, managerial and related manpower category and a 26.5% reduction in the output for skilled and semi-skilled workers. Although this output of newly trained individuals represents an inflow of trained manpower into the labour force, Jamaica lost approximately 12,370 personnel because of migration to North America (ESSJ, 1989). This represents an outflow of 60% of trained labour in a year after the country was ravaged by one of the most, if not, devastating hurricane in its history, Hurricane Gilbert, which is still talked about by those who experience the event. Having experienced such a natural disaster and losing that magnitude of trained professionals, the stage may have been set for a reduction in productivity levels. After three decades of low productivity and anaemic economic growth, and the seeming inability of successive governments and policymakers to control crime and violence, many Jamaicans often use social media to express their exit strategy or thoughts of one. The strategy predominately involves migration to escape macroeconomic deadlock in Jamaica. Many Jamaicans who work in critical or essential jobs have expressed how heartbroken they are to give the best of themselves but still feel forced to go in search of better compensation, treatment and, in their view, a life where they can comfortably afford their basic needs and some level of luxury–a better quality of life in their view. They have listened to the success stories of their colleagues who

[15] Cooper, D. (1985). Migration from Jamaica in the 1970s: Political Protest or Economic Pull? *The International Migration Review, 19*(4), 728-745. Doi: 10.2307/2546106 https://www.jstor.org/stable/2546106.

once worked alongside them and have developed the courage to chart a similar path. The recent migration of over 40 health care professionals from the University of the West Indies Hospital (UWIH) within the first two months of 2021, in the midst of COVID-19, one of the most severe health crises Jamaica has faced, is evidence of how some professionals are feeling about life in Jamaica.[16] The Minister of Health and Wellness noted that these professionals migrated to other jurisdictions to take up more attractive offers. This has resulted in a greater shortage of healthcare professionals which has placed strain on the ability of the healthcare system to deliver adequate services.

There is a shortfall of nurses in the Caribbean region, owing to "inequities in their distribution and the problem of outmigration," which has diminished access to care in many areas (Jacobson, 2015). Jacobson (2015) notes that more than 40% of nursing positions in the English-speaking Caribbean, for example, remain vacant because of migration.[17] Thompson et al. (2017) assessed factors shaping the migration of Caribbean nurses and found that in general nurses migrate for professional growth, financial benefits and educational advancement where economic factors were most commonly reported.[18] For these nurses, international migration was a "life-changing strategy".[19] At the genesis of the twenty-first century, 83% of the total stock of nurses in the Caribbean had emigrated to industrialized countries (Sands et al., 2020). While the majority of Caribbean nations experience migration of its nursing personnel, it appears that countries such as Jamaica, Haiti, Trinidad and Tobago, and Guyana experience much higher migration rates than other Caribbean countries (Sands et al., 2020). These nurses migrate for multifactorial reasons, but primarily for better remuneration (though not always the case), improved work conditions and career advancement, and professional development.

[16] Migration of nurses adding to Jamaica's COVID-19 woes. (February 24, 2021) http://radiojamaicanewsonline.com/local/migration-of-nurses-adding-to-jamaicas-covid-19-woes.

[17] Jacobson, J. (2015). The complexities of nurse migration. *AJN The American Journal of Nursing*, *115*(12), 22–23.

[18] Thompson, R., McConnell, E. S., & Corazzini, K. N. (2017). Factors shaping Caribbean nurse migration and the impact on long-term care. *Innovation in Aging*, *1*(Suppl 1), 1211.

[19] Larsen, J. A., Allan, H. T., Bryan, K., & Smith, P. (2005). Overseas nurses' motivations for working in the UK: globalization and life politics. *Work, employment and society*, *19*(2), 349–368.

Jamaica has policies that promote emigration as a poverty coping mechanism. The US markets the American dream. The UK has strong family ties, and Canada has developed policies that have become attractive to the highly skilled people worldwide. While Jamaican typically migrate to the United States and UK, recent trends have shown that there has been an increasing trend in favour of Canada relative to the United States and UK which is also evidenced in the increasing number of permanent residency cards being issued to Jamaicans since 2014. Canada lures the most qualified, skilled and talented individuals from Jamaica and other small island developing states to participate in their labour market without the burden of having to make significant investments to upgrade the already developed, qualified or trained labour. Canada has maintained high levels of immigration to offset the negative economic and fiscal impacts of its ageing population and low birth rate (El-Assal & Thevenot, 2021). Ironically, Jamaica and many other small countries are challenged with this same issue, but they are exporting human capital. As Canada intensifies its immigration through its Express Entry Program, they plan to attract over 1.2 million new immigrants between 2021 and 2023, the highest level of immigration in its history. It is also highly likely that emigration rates for university graduates in Jamaica and other small countries will increase. Canada's Express Entry Program immigration process filters people based on a combination of factors. Primary factors may include age, education level, work experience and the level of proficiency in English, French or both languages. Other factors may include experience in the Canadian labour market, education system or having adult relatives residing in Canada. It is a points-based system where the best combination normally gives a higher score and increases the chances of a person being selected to receive an Invitation To Apply (ITA) for a permanent residency card and a chance at a new life. A process like this increases the probability of emigrants making valuable contributions to the labour market as their skills can be readily and easily absorbed by the Canadian labour market and they can contribute to the productivity of the country in a meaningful and sustainable way. Jamaica and other SIDS need more friendly policies that will allow their natives to see them as the primary choice and environment to work and raise families. Or is it more beneficial if the most educated migrate and send remittances to their shores?

Appendix

See Table 6.5.

Table 6.5 Occupations of people in the Jamaican Diaspora

No.	*Occupation*	*Percent (%)*	*Cumulative (%)*	*No.*	*Occupation*	*Percent (%)*	*Cumulative (%)*
1	Teacher	17	17	31	Entrepreneur	0.7	80
2	Finance Professional	10	27	32	Factory worker	0.7	81
3	Student	10	37	33	Field supervisor	0.7	82
4	Accountant	5	42	34	Guidance Counsellor	0.7	82
5	Nurse	4	46	35	Housekeeping	0.7	83
6	Call Center Agent	3	49	36	Information Technologist	0.7	84
7	Administrative Assistant	2	51	37	Internal Audit	0.7	84
8	Police Officer	2	53	38	Machine Operator/Technician	0.7	85
9	Researcher	2	55	39	Market Researcher	0.7	86
10	Sales Man	2	58	40	Marketer	0.7	86
11	Technician	2	60	41	Marketing Representative	0.7	87
12	Attorney	1	61	42	Media Marketing Executive	0.7	88
13	Chauffeur	1	62	43	Mobile businesswoman	0.7	33
14	Cosmetologist	1	64	44	Musician	0.7	39
15	Crane operator	1	65	45	Nothing	0.7	90
16	Customer service representative	1	66	46	Office Manager	0.7	90
17	Hotel worker	1	68	47	Painter	0.7	91
18	Manager	1	69	48	Phlebotomist	0.7	92
19	Public relations manager	1	71	49	Pilot	0.7	92

(continued)

Table 6.5 (continued)

No.	*Occupation*	*Percent (%)*	*Cumulative (%)*	*No.*	*Occupation*	*Percent (%)*	*Cumulative (%)*
20	Realtor	1	12	50	Planner	0.7	93
21	Trainer	1	73	51	Planning and Research office	0.7	94
22	Business Analyst	1	74	52	Procurement Officer	0.7	95
23	Business Development	1	75	53	Psychologist	0.7	95
24	Civil Service	1	75	54	Records officer	0.7	96
25	Clerk	1	76	55	Retail representative	0.7	97
26	Contractor	1	77	56	Soldier	0.7	97
27	Data Entry Clerk	1	77	57	Supervisor	0.7	98
28	Dental assistant	1	70	53	University administrator	0.7	99
29	Economist	1	79	59	Warehouse Supervisor	0.7	99
30	Engineer	1	79	60	Web develop	0.7	100

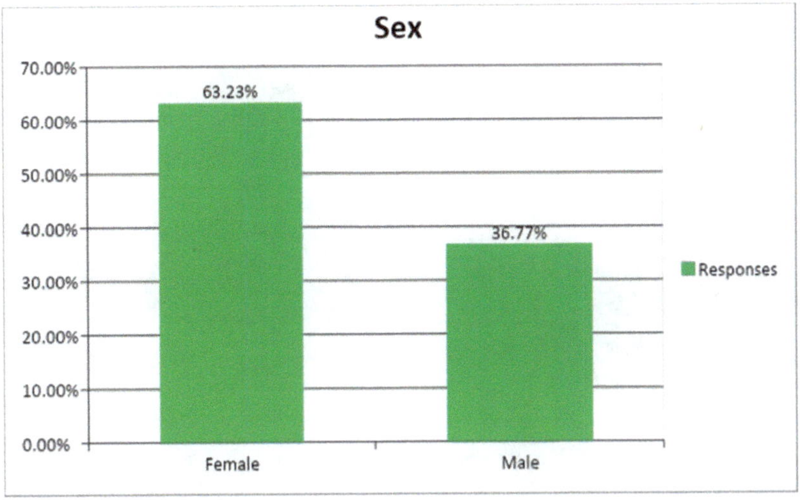

Sample selection

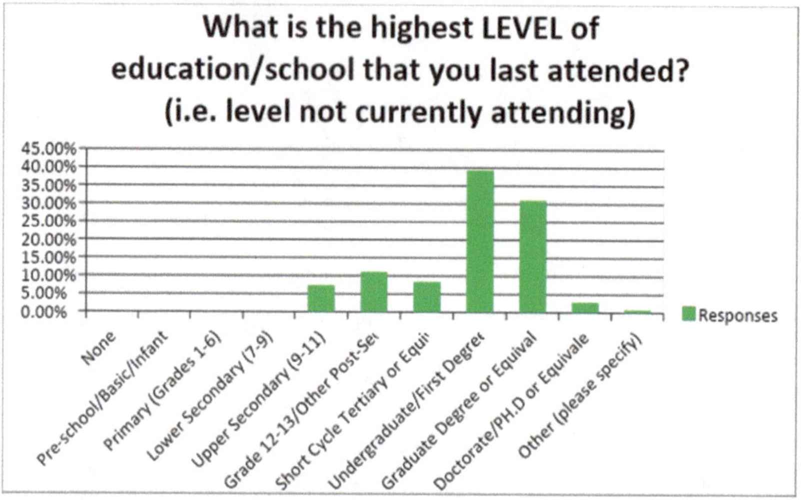
What is the highest LEVEL of education/school that you last attended? (i.e. level not currently attending)
45.00%
40.00%
35.00%
30.00%
25.00%
20.00%
15.00%
10.00%
5.00%
0.00%
None
Pre-school/Basic/Infant
Primary (Grades 1-6)
Lower Secondary (7-9)
Upper Secondary (9-11)
Grade 12-13/Other Post-Se
Short Cycle Tertiary or Equi
Undergraduate/First Degre
Graduate Degree or Equival
Doctorate/PH.D or Equivale
Other (please specify)
Responses

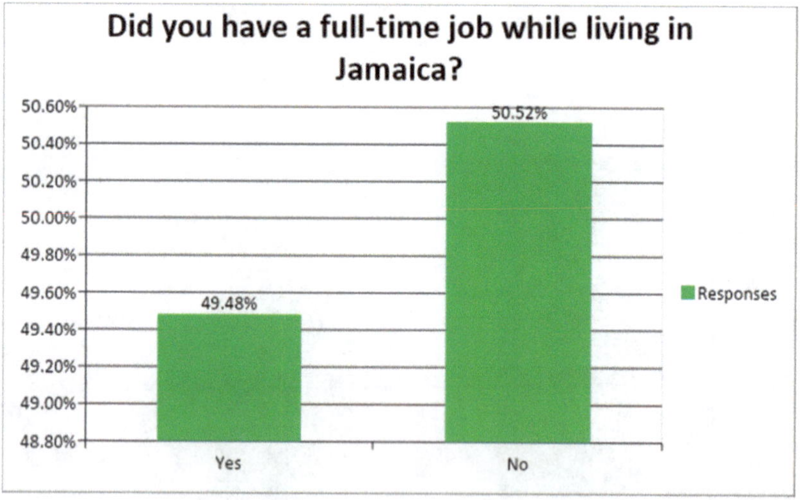
Did you have a full-time job while living in Jamaica?
50.60%
50.40%
50.20%
50.00%
49.80%
49.60%
49.40%
49.20%
49.00%
48.80%
49.48%
50.52%
Yes
No
Responses

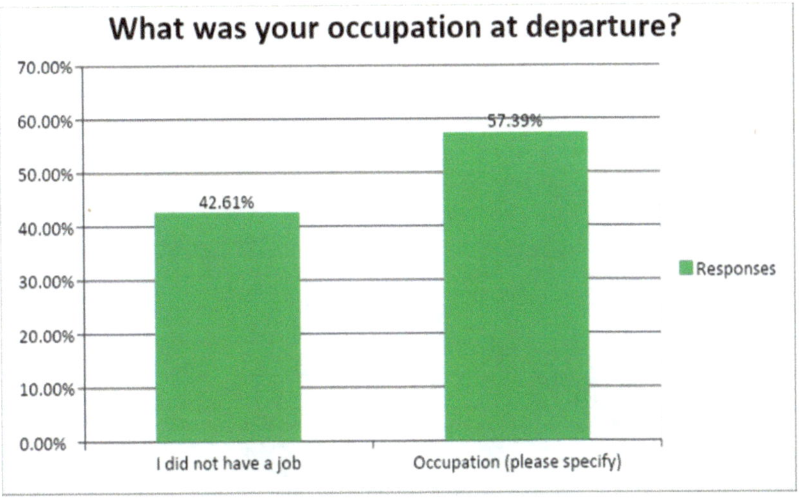
What was your occupation at departure?
70.00%
60.00%
50.00%
40.00%
30.00%
20.00%
10.00%
0.00%
42.61%
57.39%
I did not have a job
Occupation (please specify)
Responses

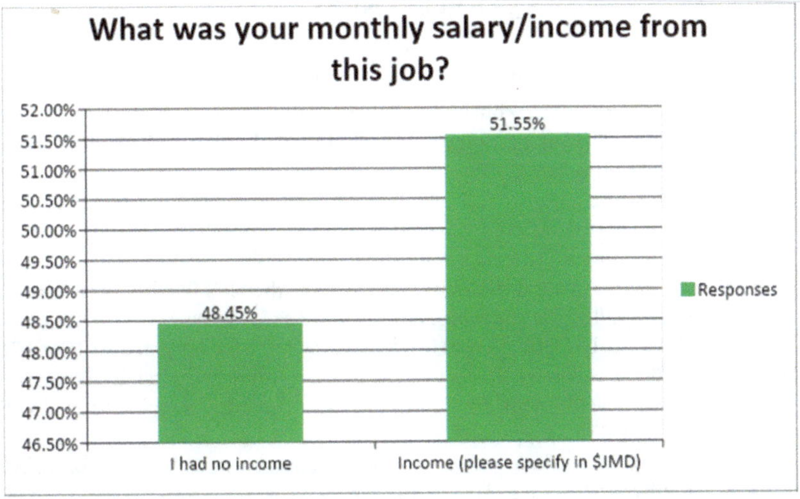
What was your monthly salary/income from this job?
52.00%
51.50%
51.00%
50.50%
50.00%
49.50%
49.00%
48.50%
48.00%
47.50%
47.00%
46.50%
48.45%
51.55%
I had no income
Income (please specify in $JMD)
Responses

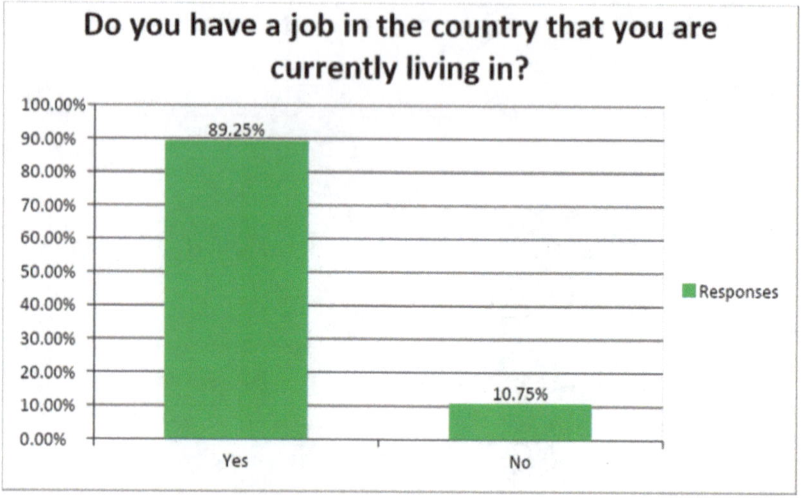

References

Artuc, E., Docquier, F., Özden, Ç., & Parsons, C. (2014). *A global assessment of human capital mobility: The role of non-OECD destinations*. The World Bank. https://www.econstor.eu/bitstream/10419/107527/1/dp8746.pdf

Castelli, F. (2018). Drivers of migration: Why do people move?. *Journal of travel medicine, 25*(1), tay040.

Chung, S., Sudo, R., Mack, P. J., Wan, C. R., Vickerman, V., & Kamm, R. D. (2009). Cell migration into scaffolds under co-culture conditions in a microfluidic platform. *Lab on a Chip, 9*(2), 269–275.

Cooper, D. (1985). Migration from Jamaica in the 1970s: Political protest or economic pull? *The International Migration Review, 19*(4), 728–745. https://doi.org/10.2307/2546106, https://www.jstor.org/stable/2546106

Di Maria, C., & Stryszowski, P. (2006). Migration, human capital accumulation and economic development, http://citeseerx.ist.psu.edu/viewdoc/download?doi=10.1.1.514.3533&rep=rep1&type=pdf

Docquier, F. The brain drain from developing countries. IZA World of Labor 2014: 31, https://doi.org/10.15185/izawol.31

Docquier, F., & Marfouk, A. (2004). *Measuring the international mobility of skilled workers (1990–2000): Release 1.0*. The World Bank. https://openknowledge.worldbank.org/bitstream/handle/10986/14126/wps3381.pdf?sequence=1

Docquier, F., & Schiff, M. (March, 2008). *Measuring skilled emigration rates: The case of small states*. http://ftp.iza.org/dp3388.pdf

El-Assal, K., & Thevenot, S. (2021). Canada breaks all-time immigration record by landing 401,000 immigrants in 2021. Cic News, 23.

Haque, N. U., & Kim, S. J. (1995). "Human capital flight": Impact of migration on income and growth. *Staff Papers, 42*(3), 577–607.

International Organization of Migration. (2019). World Migration Report 2020. https://publications.iom.int/system/files/pdf/wmr_2020.pdf

Jacobson, J. (2015). The complexities of nurse migration. *AJN the American Journal of Nursing, 115*(12), 22–23.

Larsen, J. A., Allan, H. T., Bryan, K., & Smith, P. (2005). Overseas nurses' motivations for working in the UK: Globalization and life politics. *Work, Employment and Society, 19*(2), 349–368.

Liao, P. J., Wang, P., Wang, Y. C., & Yip, C. K. (2017). Educational choice, rural-urban migration and economic development (No. w23939). *National Bureau of Economic Research.*

National Immigration Forum. (July, 2019). *Push or pull factors: What drives Central American Migrants to the U.S.?* https://immigrationforum.org/article/push-or-pull-factors-what-drives-central-american-migrants-to-the-u-s/#:~:text=Push%E2%80%9D%20factors%20are%20conditions%20in,factors%20include%20violence%2C%20gender%20inequality%2C

Rolle Sands, S., Ingraham, K., & Salami, B. O. (2020). Caribbean nurse migration—a scoping review. *Human Resources for Health, 18*(1), 1–10.

Thet, K. K. (2014). Pull and push factors of migration: A case study in the urban Area of Monywa Township, Myanmar. *News from the World of Statistics, 1*(24), 1–14.

Thompson, R., McConnell, E. S., & Corazzini, K. N. (2017). Factors shaping Caribbean nurse migration and the impact on long-term care. *Innovation in Aging, 1*(Suppl. 1), 1211.

Wenner, M. (September, 2016) Brain drain: A curse of small states? https://blogs.iadb.org/caribbean-dev-trends/en/brain-drain-a-curse-of-small-states/

Younas, M. Z., & Ahmad, G. (2020). Examining the impacts of human capital flight on economic growth: A case study of Pakistan. *Review of Knowledge Economy, 7*(1), 1–30.

CHAPTER 7

Remittances

7.1 Remittance Benefits and Overview of Small Countries

Human capital flight contributes positively to the income of their home countries through the transfer of remittances. The benevolence of people who have either migrated permanently or temporarily who send money back home. As at 2019, worldwide remittances received surpassed US$650 billion and has climbed to about 76% of global GDP (See Fig. 7.1). In addition, remittances received by small states have increased by 362% between 1990 and 2019.[1] The surge in remittances has been due to the sizeable depreciation of the US dollar against most other major currencies, especially the Euro which has increased the value of non-dollar remittance over time, and improvements in data recording methods of Central Banks.[2] Growing migration has also played a role that will continue to influence remittance growth as people continue to leave their home countries. Remittances have been identified as the third pillar of development as the volume is second to foreign direct investment and

[1] Authors' calculation based on Data retrieved from the World Development Indicators, 2021.

[2] World Bank. (2005). *Global development finance 2005: Mobilizing finance and managing vulnerability, volume 1. Analysis and statistical appendix. Global development finance* . World Bank. https://openknowledge.worldbank.org/handle/10986/8135 License: CC BY 3.0 IGO. http://hdl.handle.net/10986/8135.

A. Haughton and W. Ivey, *Overcoming Productivity Challenges in Small Countries*, https://doi.org/10.1007/978-3-031-23301-2_7

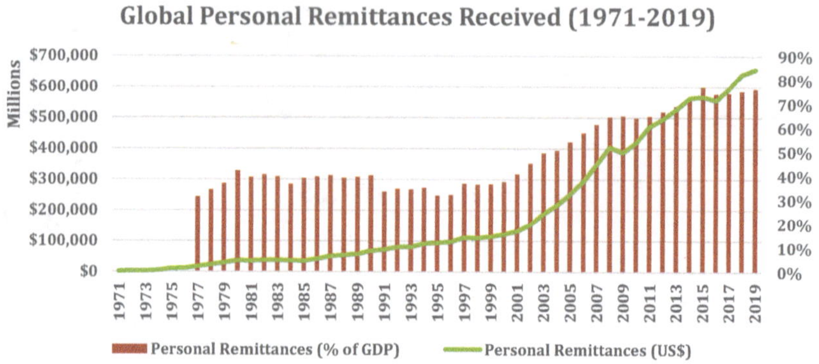

Fig. 7.1 Global remittances received (*Source* World Bank)

higher than overseas development assistance (Alfieri et al., 2006). Temporary cases may include farm workers and students engaging in work and travel arrangements to complete their university education. Remittances are critical inflows that add to a country's balance of payments.

Remittances are defined as the sum of workers remittances, compensation of employees and others include migrants' transfers in addition to the abovementioned.[3] In this chapter, remittances are defined as the sum of personal transfers and compensation of employees, as defined in the sixth edition of the IMF's Balance of Payments Manual.[4] Firstly, personal transfers are classified as all current transfers in cash or kind made or received by resident households to or from nonresident households. Therefore, personal transfers include all current transfers between resident and nonresident individuals. Secondly, compensation of employees refers to the income of border, seasonal, and other short-term workers who are employed in an economy where they are not residents and of

[3] Alfieri, A., Havinga, I., & Hvidsten, V. (2006, February). Issue paper: Definition of remittances. In *Meeting of the United Nations technical subgroup on movement of persons–mode* (Vol. 4). https://millenniumindicators.un.org/unsd/tradeserv/TFSITS/TSGMode4/tsg3-feb06/tsg0602-14.pdf.

[4] Balance of Payments Manual, Sixth Edition Compilation Guide. https://books.google.com.jm/books?id=q0bnBwAAQBAJ&printsec=frontcover&dq=Balance+of+Payments+Manual,+Sixth+Edition+Compilation+Guide&hl=en&sa=X&redir_esc=y#v=onepage&q=Balance%20of%20Payments%20Manual%2C%20Sixth%20Edition%20Compilation%20Guide&f=false.

residents employed by nonresident entities. Over time, remittances have become an increasingly important source of foreign exchange inflows in many developing countries. Receiving remittances is not a random event as households that receive remittances are likely to exhibit certain characteristics, such as having family members abroad, or specific needs owing to their composition (perhaps they have more dependent children or elderly members) (Amuedo-Dorantes, 2014).[5]

7.2 Remittance Inflows by Region

7.2.1 South Asia

Across the world, South Asian countries have been receiving the greatest level of remittance as a percentage of GDP (see Fig. 7.2). This is a consistent trend that began over four decades ago. The region is made up of 8 countries: Afghanistan, Bangladesh, Bhutan, India, Maldives, Nepal, Pakistan and Sri Lanka. Based on the World Bank's country classification, all countries within this region except for the Maldives fall in the lower-middle-income category or lower. In other words, they have a maximum Gross National Income (GNI) per capita of US $4095.[6] According to the World Bank, South Asian countries alone account for about 29% of extreme global poverty.[7] This partially explains the high level of remittances received from their diaspora. The past 3 decades have seen a significant elevation in remittance inflows in the region. Remittance inflows have grown from 1.5% of GDP in 1989 to 3.9% in 2019. However, flows have experienced different shocks within the period. In 2003, there was a positive shock to remittances in the region. This was particularly reflected in the receipts from Pakistan which rose due to concerns and heightened scrutiny by immigration authorities in developed countries that influenced outward remittance of savings by undocumented

[5] Amuedo-Dorantes, C. (2014). *The good and the bad in remittance flows*. IZA World of Labor. https://wol.iza.org/articles/good-and-bad-in-remittance-flows/long.

[6] World Bank Country and Lending Groups. https://datahelpdesk.worldbank.org/knowledgebase/articles/906519-world-bank-country-and-lending-groups.

[7] https://worldbank.github.io/SARMD_guidelines/poverty-measures.html.

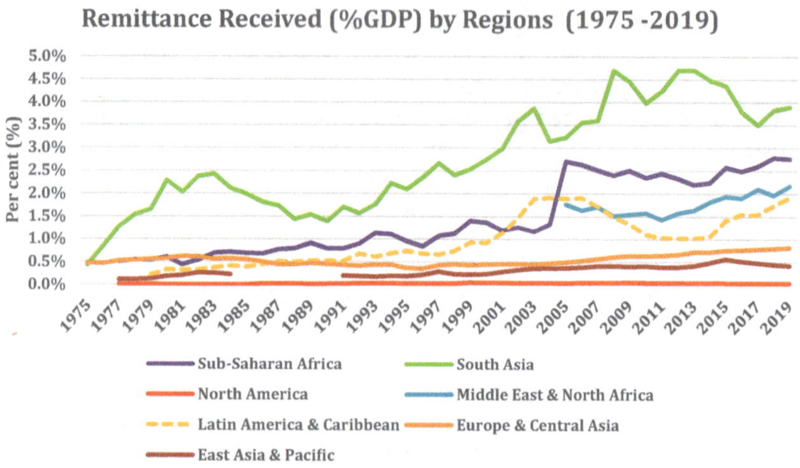

Fig. 7.2 Remittance received by regions (*Source* World Bank)

migrants.[8] Between 2001 and 2003, this culminated in a tripling of remittance flows in Pakistan and the region by extension. On the contrary, the Global Financial Crisis induced a negative shock on remittance flows to the region. Remittances fell by approximately 0.8 percentage points (pps) from a peaked of 4.7% in 2008 to 3.9% in 2010. As the crises dissipated, inflows regained its pre-crisis momentum before taking a nosedive again towards 2017.

7.2.2 *Sub-Saharan Africa*

Following the South Asian region is the Sub-Saharan Africa region which receives the second highest level of remittances in the world. The level of remittance received in this region has trended below 1.5% up until 2005 when receipts were valued at about 2.7% of GDP. In this same year,

[8] World Bank. (2005). *Global development finance 2005: Mobilizing finance and managing vulnerability, volume 1. Analysis and statistical appendix. Global development finance* . World Bank. https://openknowledge.worldbank.org/handle/10986/8135. License: CC BY 3.0 IGO. http://hdl.handle.net/10986/8135.

remittance flows were estimated to have surpassed $233 billion worldwide which developing countries received 167 billion.[9] In this context, developing countries received 71.7% of global remittances. This period also coincided with a 107.7% increase in remittance inflows to the region which was the sharpest historical increase in inflows to Sub-Saharan Africa and of any other region during the time (see Fig. 7.2). After 2005, remittance as per cent of GDP tapered off in the subsequent years to approximately 2.2% in 2013 before venturing on another positive incident of growth up to 2019. Ajayi et al. (2009)'s research shows that remittances have to some extent contributed to the improvement of well-being in sub-Saharan Africa. If this source of foreign exchange is expected to consistently have a material impact on the well-being of the people of sub-Saharan Africa, Ajayi et al. (2009) propose that greater effort must be placed on macroeconomic policy, flexible financial sector infrastructure and low transaction cost of fund transfers, conducive investment climate with a stable political environment, adequate checks on the incidence of money laundering, greater transparency in the transfer of fund between the sender, the disbursing agent and the receiver, adequate protection of the customers, financial security and risk management practices were suggested.[10]

7.2.3 *Latin America and the Caribbean*

The Latin American and Caribbean region receives the third highest level of remittance in the world. The region has remained consistently at this position relative to the other region. In this region, approximately 80% of remittance comes from the United States. Over time inflows in the region has gradually increased. Of note, in the late 1990s and early 2000s, remittance increased sharply despite a recession in the United States as well as the September 11 terrorist attacks of 2001. This positive trajectory was short-lived as the GFC induced a decade long slump in remittance

[9] World Bank. (2006). *Global development finance 2006: The development potential of surging capital flows, Volume 1. Review, analysis, and outlook. Global development finance*. World Bank. https://openknowledge.worldbank.org/handle/10986/8125 License: CC BY 3.0 IGO. http://hdl.handle.net/10986/8125.

[10] Ajayi, M. A., Ijaiya, M. A., Ijaiya, G. T., Bello, R. A., Ijaiya, M. A., & Adeyemi, S. L. (2009). International remittances and well-being in sub-Saharan Africa. *Journal of Economics and International Finance*, *1*(3), 78–84. https://academicjournals.org/journal/JEIF/article-full-text-pdf/F7388961928.

as a share of GDP which recovered in 2019. We will decompose this region into Latin American and Caribbean countries and discuss selected Caribbean countries in more detail in the next section.

7.2.3.1 Caribbean Countries

Remittances are a very important source of foreign exchange for small countries including many from the Caribbean region. Based on data compiled by the World Bank, remittance as a percentage of GDP has increased by 1.7 percentage points from 0.2% in 1979 to 1.9% in 2019. Over the period, there have been fluctuations in both the magnitude and direction of inflows emanating from economic crises and policy decisions (see Fig. 7.3). Inflows to the region increased steadily from 0.2% in 1979 to 0.9% up until 1999. Thereafter, it increased sharply before plateauing at about 1.9% in 2003. This preceded the Global Financial Crisis of 2007. The Financial Crisis induced a downward spiral that lasted for approximately 5 years before it levelled off at about 1% in 2011. Inflows to the region picked up again in 2014 and increased by 0.9% points by the end of 2019.

While remittance is a useful and critical source of foreign exchange that small countries can benefit from, some countries receive way more remittance than others. The structural make-up of these economies may be a driving factor behind these disparities. For example, between 1999 and 2019, countries such as Belize (4.6%), Barbados (3.1%), St. Lucia (1.9%), Antigua and Barbuda (1.9%), Suriname (0.7%) and Trinidad and Tobago (0.6%) averaged remittance inflows of less than 5% of their GDP. On the other hand, countries such as St. Vincent (5%) Dominican Republic (7.9%), Guyana (9.3%), Haiti, (14%) and Jamaica (14.8%) have averaged at least 5% per annum. By assessing the trend of remittance inflows in the top 8 remittance receivers in the Caribbean, there is evidence to suggest that, on average Jamaica receives the largest level of remittances relative to its economic output compared to any other Caribbean country. The island has maintained this position for almost every year except for the times when there is a spike in Haiti and Guyana. However, since 2017 Haiti has become the front runner in remittance in the Caribbean by receiving on average 20.7% of GDP (See Fig. 7.3).[11] There has been a noticeable upswing in Haiti since the catastrophic earthquake in 2010

[11] In plotting Haiti's trendline, the average of 6.2% for 1971 to 1986 is used to plot the graph points between 1990 to 1997 due to data unavailability for those years.

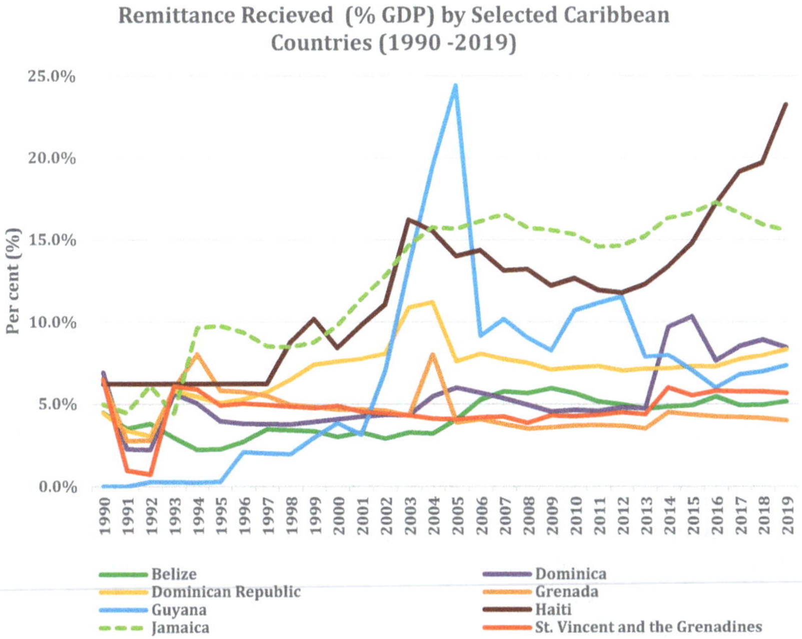

Fig. 7.3 Remittance received in the Caribbean (*Source* World Bank)

that crippled the Haitian economy. The earthquake compounded the effects of a country that was battered by tropical storm Fay and hurricanes Gustav and Hanna in 2008. With the loss of much of its human capital and damage to Haiti's economic infrastructure, a coping mechanism was needed. One strategy that Haitians have used over the decades is migration. Many Haitians view migration as a survival strategy for individual migrants and their families to now buttress the Haitian economy through the transfer of remittances.[12] There will likely be a further spike in remittances to the island as it has been hit hard in 2021 by the COVID-19 crisis, a 7.2 magnitude earthquake in August which left over 650,000

[12] Cela, T. (2020, May). *Haitian families and loss of remittances during COVID-19 Pandemic.* https://oecd-development-matters.org/2020/05/29/haitian-families-and-loss-of-remittances-during-the-covid-19-pandemic/

people needing assistance and have been impacted negatively by tropical storm Grace. Roberts (2006) notes that the impact of remittances on most Caribbean economies is positive. However, the inflows have a stronger impact on Haiti, Jamaica and Guyana.[13] He reasoned that because Caribbean economies have a history of severe poverty, in the absence of these additional funds, the poverty level for the recipients' households would have been significantly higher in absolute terms and terms severity. In other words, remittance is a safety net for small countries in the Caribbean.

7.2.4 *Other Regions*

The other three regions, Europe and Central Asia, East Asia and the Pacific, and North America do not receive the level of remittances as the other four regions. This is because in most cases, the countries within these regions make up a large proportion of the diaspora of poorer countries. Hence, remittance received as a share of GDP is lower than other regions. This is a plausible outcome as these regions are where people normally migrate to. As a result, remittances tend to flow from these regions rather than to them.

7.3 Remittance Flows to Jamaica

Remittance inflows as a share of GDP have gradually been increasing over the last 3 decades and have become an integral part of the livelihoods of many Jamaicans. It has grown from 5% in 1990 to 15.6% in 2019 (see Fig. 7.4). Over the last decade[14] (January 2012–March 2021), the remittance by corridor inflows to Jamaica have exceeded US$20 billion. Inflows have increased by 199% between January 2012 and March 2021. The COVID-19 pandemic has induced a growth in remittance received in Jamaica that is unprecedented. People within the Jamaican diaspora have poured money intensively into Jamaican households to support those who have lost their jobs and those whose incomes have been interrupted by the

[13] Roberts, D. (2006). *The development impact of remittances on Caribbean economies: The case of Guyana*. In SALISES Conference, University of the West Indies, St. Augustine. https://sta.uwi.edu/conferences/salises/documents/roberts%20d.pdf.

[14] Reference Period is January 2012–March 2021(1 year after the first COVID-19 case was recorded in Jamaica).

fiscal measures put in place to contain the spread of the virus and limited movement. From the onset of the pandemic in March 2020–March 2021, a year later, about US$52.5 million have flowed to Jamaican households. This represents a 65% elevation in remittances from US$275 million to US$327.5 million (see Fig. 7.5).

Inflows have come mainly from the United States (US$11.7 billion), the United Kingdom (2.5 billion), Canada (2 billion) and the Cayman

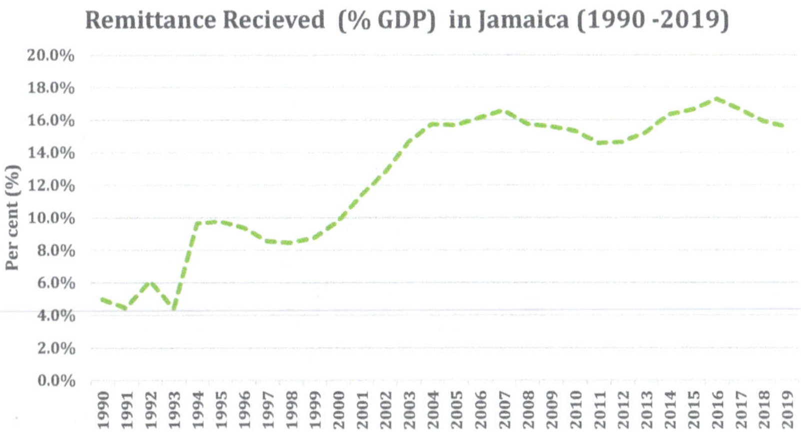

Fig. 7.4 Remittances received (%GDP) (*Source* World Bank)

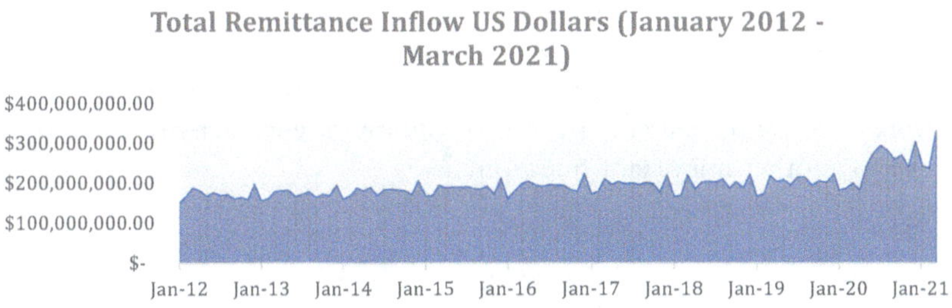

Fig. 7.5 Total remittances inflows (*Source* Bank of Jamaica)

Islands (1.1 billion). Based on a remittance survey[15] conducted by the Bank of Jamaica in 2010, remittances are predominately used for consumption purposes in Jamaica. Approximately, 85% is spent on food, education, utility bills and other basic living expenses; 9% is saved and only 6% goes towards investment decisions. The survey also gave two important insights which are crucial in understanding the remittance and labour market dynamics in Jamaica. Firstly, remittances are typically sent to people who are between the ages of 26 and 40 years old. This age group can be considered as the prime working-age range. On one hand, people within this age group are likely to be at an advanced level in their development training. This could be up to the graduate level in the context of university or up to an intermediate or expert level for vocational skills training from an institution such as HEART or other similar institutions. Secondly, approximately 67% of remittance receivers actively participated in the labour force.

One would expect that remittance recipients would largely be people who are unemployed or do not have the means to earn an income, especially since remittances are used for basic living expenses. Is it that the income that these persons are earning is so low that it must be supplemented by other sources of income? It is widely accepted on the island that minimum wage and income levels, in general, are not competitive nor sufficient to sustain a decent quality of life to the extent where the government gave life support to the disposable incomes of people who earned below a certain level in 2016. Ivey (2017)[16] estimated the impact of minimum wage increases on Quality of Life on 107 countries worldwide. Ivey (2017) found that while increases in minimum wage marginally improve the quality of life, it is not sufficient, but many people are still enjoying a decent quality of life. Ivey (2017) attributed this to other factors such as remittances, informal economy or elicit activities. We now move to another important question.

[15] Ramocan, E. G. (2010). *Remittances to Jamaica. Findings from a national survey of remittance recipients*. Bank of Jamaica https://www.boj.org.jm/uploads/pdf/papers_pamphlets/papers_pamphlets_Remittances_to_Jamaica_-_Findings_from_a_National_Survey_of_Remittance_Recipients.pdf.

[16] Ivey, W. (2017). An empirical analysis of minimum wage on the quality of Life. *Southwestern Journal of Economics, 12* (1).

7.4 The Remittance and Productivity Dilemma

What happens to the output of a country when the income earned from informal economy activities or income received from remittances outweighs the income from participating in the labour market? People respond to incentives, and they are going to choose the outcomes or options which maximize their well-being. However, this may not be the outcome that maximizes the total welfare of the country. For example, if a student decides that the remittances that he or she is receiving from his or her parent affords a decent standard of living relative to their peers and they can sport the latest clothes, shoes (clarkes or bridgets), gadgets, eat at the finest restaurants on the island and attend the most exotic parties, given that we have already established that income levels are not satisfactory, how will this person be motivated to join the labour force especially if compensation from that job is lower than the remittances? This may not be the case for all individuals who receive remittances as some receivers use remittances as a vehicle to evade poverty by paying for university especially if they did not qualify for a student's loan or were just fortunate enough to go through school without having to consider a loan. What about people who are working full-time jobs and receive remittances which exceed their gross income? Will these persons go above and beyond the call of duty? What role do these scenarios play in the low and declining output per worker of Jamaica? Are people substituting effort in formal jobs with remittances and other sources of income and as a result output has average the same growth rate for over 30 years? In Section 7.5, we estimate an ARDL cointegration model to analyse the short-run dynamics and the long-run relationship between remittance and output growth in Jamaica. Before we do that, let us briefly look at how remittance inflows have been affected in different economic crises.

7.4.1 Global Financial Crisis and COVID-19 Impact on Remittances to Jamaica

In an economic crisis or recession, central banks typically employ tools within its monetary policy toolbox to navigate the turbulence. Normally, the central bank of the country would try to stimulate its economy by lowering its monetary policy rate. In the case of Jamaica, the monetary policy rate is the interest rate on overnight balances in the current accounts of deposit taking institutions held at the Bank of Jamaica. The policy rate of 0.5% during the initial phases of the pandemic left little room for interest rate to fall further and be used as a tool to stimulate Aggregate Demand (AD) through more borrowing for investment spending. The severity of recession too, will require more robust than normal open market operations (OMO). When this happens, the central bank may have to employ more unconventional tools if it is going to give itself a chance to change the trajectory of the economy especially if the policy rate is already at or close to zero. itative Easing (QE) may be one of the preferred unconventional choices. This is a monetary policy tool that was formulated by Richard Werner, a German economist, in relation to the Japanese crisis in 1995. QE is an unconventional monetary policy tool where the central bank purchases longer-term securities which may include government bonds or mortgage-backed securities from the open market to increase the money supply and encourage lending and investments.[17] This effectively increases the central bank's asset base, puts money into the economy, reduces interest rates and stimulates AD. Based on the measure selected from the monetary policy toolkit to stimulate the economies of rich countries especially in a global pandemic, it may result in different levels of remittance outflows to small countries like Jamaica.

According to the BOJ's 2010 survey, approximately 70% of respondents related that their remittance senders were negatively affected by the GFC. They noted that the numbers of hours in senders worked were reduced, many lost their place in the employed labour force, and some also faced bills and mortgage dilemmas. This culminated in a reduction in the frequency as well as the amount of money that was remitted to receivers. More than half of Jamaican remittance receivers continued to collect at a lower frequency and took home a lower amount than they were accustomed to while 39% received at their normal frequencies. Based on data from the World Bank, for the entire period (2007–2009) of the crisis, remittance inflows to Jamaica contracted by about 11%. Fast forward to the COVID-19 crisis, which was induced by the outbreak of the coronavirus, remittance inflows by corridors to Jamaica reached historic highs. Jamaica confirmed its first case in March of 2020 and two months later, remittance inflows increased by about 16%, and by June 2020 it increased to 42% year over year. By March 2021, one year after

[17] https://www.investopedia.com/terms/q/itative-easing.asp.

Jamaica recorded its first case, remittance was up by 65%. Massive stimulus packages in the United States, Canada and the UK contributed to the increase in remittances sent to Jamaica, a stark contrast to the GFC of 2008. Why was this so? Why were remittances negatively affected by the crisis in 2008, but positively affected by the crisis in 2020? Both crises harmed income and employment, however, the channel through which the policy was administered was very different. In response to the GFC of 2008/2009, the Federal Reserve used QE to increase its balance sheet.[18] From December 2008 to March 2010, the Fed purchased 175 million in agency securities and $1.25 trillion in mortgage-backed securities. QE was exercised in many other countries including Switzerland, the UK, Sweden and other countries within the Euro Area. The monetary instrument flowed to the banks with little or no implication for the income of the employed. On the contrary, in response to the 2020 coronavirus, the United States injected a $2 trillion stimulus package. The largest in its history, which flowed directly to the working class instead of going to the banks and financial institutions. Some of this money managed to leak from the US economy through formal remittance channels as people living within the diaspora sent money back home to help family, friends and loved ones who lost their incomes to navigate the pandemic. Remittances also flowed to small countries through the accounts of former work and travel employees who still had active bank accounts that were opened in the United States. This may partly explain the difference between remittance inflows in both crises. An article[19] by the Wall Street Journal showed that households in the United States added approximately 13.5 trillion in wealth in 2020. This represents the most profound increase in wealth over the last 30 years. Americans were able to reduce their credit card debt, increase their savings and refinance their mortgage and benefit from cheaper interest rates. Underpinning this increase in wealth was the stock market. The S&P 500 grew by 29% a year after the recession started which accounted for more than 50% of the growth in wealth. Based on projections made by JPMorgan and Chase Institute, the checking accounts of the bottom 25% of lower-income earners had risen by about 50% relative to the previous year. The source of income in these accounts

[18] https://www.stlouisfed.org/publications/regional-economist/third-quarter-2017/quantitative-easing-how-well-does-this-tool-work.

[19] https://www.wsj.com/articles/during-covid-19-most-americans-got-richerespecially-the-rich-11624791602.

was derived from stimulus checks and unemployment benefits that the government used to buttress the impact of the pandemic. On the other hand, in 2008 households in the United States lost about 8 trillion dollars. While the stock market grew in 2020, it had declined by 38.49% in the global financial crisis (2007–2009).

Monetary and fiscal policy positions in larger countries have implications on the remittances of small countries like Jamaica. This is especially true as it relates to the United States since it's through that corridor that Jamaica receives most of its inflows. This is especially important when the world economy or a part of it is in crisis mode. With globalization and the increased interlocking or interconnectedness of countries, small countries must pay attention to these policies and assess the impact they may have on their economies. While some policies in a crisis or recession can have positive spillovers in the case of COVID-19 and remittances others can have deleterious effects as with the GFC and reduced remittance inflows.

7.5 The Long-Run Relationship Between GDP Growth and Remittances

Remittance inflows can have both positive and negative implications for the efficient functioning of a nation's economy. On a positive note, remittance inflows can act as a buffer against adverse economic shocks at both the country and household levels. It can also increase the well-being of receiving households by smoothing consumption and improving living conditions. It can facilitate human capital enhancement by providing the income to improve sanitary conditions, healthier lifestyles, proper health care and greater educational attainment. Remittances can assuage the credit constraints of the unbanked households in poor rural areas, facilitate asset accumulation and business investments, promote financial literacy and reduce poverty (Amuedo-Dorantes, 2014; Hosny, 2020). Results emanating from Husien (2019) suggest that remittances have a positive and significant long-run impact on economic growth. On the flip side, remittances may influence people to become lackadaisical, reduce their participation in the labour force, consume conspicuously which are recipes for a reduction in output per worker and ultimately poverty acceleration. Chami et al. (2005)[20] found evidence to suggest that remittances may reduce the recipients' motivation to work which impairs their personal productivity. To this end, remittances can reduce labour supply and create a culture of dependency that inhibits economic growth, increase the consumption of non-tradable goods, raise their prices, appreciate the real exchange rate and decrease exports, thus damaging the receiving country's competitiveness in the world markets (Amuedo-Dorantes, 2014). To remedy or avert these adverse effects of remittances, countries can apply anti-immigrant measures to restrict the movement of people to more developed host countries like the United States, Europe and Asia that are attractive for migration. The question is what is the impact of remittances inflows on the Jamaican economy?

Although remittances have their positives and negatives regarding economic growth, it is a useful poverty alleviation tool. Imran et al. (2020)[21] studied the relationship between remittances and the incidence and severity of poverty in the Punjab province of Pakistan. They controlled for several factors including the number of dependents in households, age, gender and the education level of the head of households. The results indicated that remittance inflows reduce the incidence and severity of poverty. Azizi (2021)[22] also investigated the impact of remittances on poverty and inequality by using data from 103 developing countries over a 2-decade horizon. Similar to Imran et al. (2020), Azizi (2021) found that remittances decrease poverty. More specifically, Azizi (2021) found that a 10% increase in per capita remittances leads to a 1% decrease in poverty headcount and a 1.8% decrease in the poverty gap. Auxiliary findings emerging from Azizi (2021) provides evidence that remittances decrease inequality in developing countries. In other words, remittances increase the income share attributed to the poorest decile

[20] Chami, R., Fullenkamp, C., & Jahjah, S. (2005). Are immigrant remittance flows a source of capital for development. *IMF Staff Papers, 52*(1), 55–81.

[21] Imran, K., Devadason, E. S., & Kee Cheok, C. (2020). Foreign remittances and regional poverty: Evidence from household data. *International Migration*, *58*(4), 214–230. https://www.researchgate.net/profile/Kashif-Imran4/publication/338009308_Foreign_Remittances_and_Regional_Poverty_Evidence_from_Household_Data/links/5e0c2197a6fdcc28374d4254/Foreign-Remittances-and-Regional-Poverty-Evidence-from-Household-Data.pdf.

[22] Azizi, S. (2021). The impacts of workers' remittances on poverty and inequality in developing countries. *Empirical Economics*, *60*(2), 969–991. https://link.springer.com/article/10.1007/s00181-019-01764-8.

Table 7.1 Summary statistics

	GDPG	*REM*	*CPS*	*GXP*	*INFLATION*	*POPG*	*GCF*	*TRADE*
Mean	1.0163	9.7291	26.1844	15.0413	15.7247	0.8504	22.8189	91.5819
Median	1.1857	9.4557	25.9850	14.2546	9.9403	0.7395	23.2169	89.4156
Max	9.4171	17.2864	42.9274	21.9529	77.2966	1.6765	29.8286	125.4943
Min	−6.6942	2.3264	12.9521	9.3119	2.3510	0.4564	12.2262	58.1491
Std. Dev	3.1029	5.4578	6.1762	3.0577	14.0143	0.3584	3.6230	13.6129
Obs	44	44	44	44	44	44	44	44

and quintile and decrease the income share held by the richest quintile and decile in developing countries. Adeseye (2021)[23] studied the effect of remittance flows on the Nigerian economy using annual data over 29 years (1990–2018). The study revealed that there exists a significant relationship between remittances, GDP growth and Trade (exports and imports) in Nigeria. It will be interesting to see what the analysis will reveal about the remittance impact on the Jamaican economy.

7.5.1 *Data and Methodology*

7.5.1.1 Data

The data employed in this study was collected from the World Development Indicators provided by the World Bank. Annual data was collected for 1976 to 2019 (44 years inclusive). The variables collected are GDP growth (GDPG) and Population Growth (POPG), which are measured as the changes in GDP and Population in percentage; Inflation (INF), measured as the percentage change in the consumer price index; Personal Remittances received (REM), Domestic Credit to private sector (CPS), General Government final consumption expenditure (GXP), Gross Capital Formation (GCF) and Trade (TRADE) which is the sum of imports and exports. All these variables are measured as a percentage or share of GDP and their summary statistics are produced in Table 7.1.

[23] Adeseye, A. (2021). The effect of migrants remittance on economy growth in Nigeria: An Empirical Study. *Open Journal of Political Science*, *11*(01), 99.

7.5.2 Methodology

We estimate the relationship between GDP growth and Remittance in Jamaica by following a similar model employed by Anetor (2019) who analysed the economic growth and remittance nexus in Nigeria. The model is as follows: The ARDL cointegration approach was developed by Pesaran and Shin (1999) and Pesaran et al. (2001). It has three advantages in comparison with other previous and traditional cointegration methods. The first one is that the ARDL does not need that all the variables under study must be integrated of the same order and it can be applied when the underlying variables are integrated of order one, order zero or fractionally integrated. The second advantage is that the ARDL test is relatively more efficient in the case of small and finite sample data sizes. The last and third advantage is that by applying the ARDL technique, we obtain unbiased estimates of the long-run model (Harris & Sollis, 2003).

7.5.2.1 Model Specification

$$\begin{aligned}\Delta \mathrm{GDPG}_t = \beta_0 + \beta_1 \mathrm{REM}_t + \beta_2 \mathrm{CPS}_t + \beta_3 \mathrm{GXP}_t + \beta_4 \mathrm{INF}_t \\ + \beta_5 \mathrm{POPG}_t + \beta_6 \mathrm{GCF}_t + \beta_7 \mathrm{TRADE}_t + \epsilon_t\end{aligned} \tag{7.1}$$

where REM = remittances, CPS = credit to the private sector, INF = inflation, POPG = population growth, GCF = Government Final Expenditure and TRADE = the sum of imports and exports as a percentage of GDP.

7.5.2.2 Auto-Regressive Distributed Lag Model

$$\begin{aligned}\mathrm{GDPG}_t = \beta_0 &+ w_1 \mathrm{GDPG}_{t-1} + w_2 \mathrm{REM}_{t-1} + w_3 \mathrm{CPS}_{t-1} + w_4 \mathrm{GXP}_{t-1} \\ &+ w_5 \mathrm{INF}_{t-1} + w_6 \mathrm{POPG}_{t-1} + w_7 \mathrm{GCF}_{t-1} + w_8 \mathrm{TRADE}_{t-1} \\ &+ \sum_{i=1}^{n} \beta_i \mathrm{GDPG}_{t-1} + \sum_{i=1}^{n} \alpha_i \mathrm{REM}_{t-1} + \sum_{i=1}^{n} \delta_i \mathrm{CPS}_{t-1} \\ &+ \sum_{i=1}^{n} \varphi_i \mathrm{GXP}_{t-1} + \sum_{i=1}^{n} \gamma_i \mathrm{INF}_{t-1} + \sum_{i=1}^{n} \lambda_i \mathrm{POPG}_{t-1} \\ &+ \sum_{i=1}^{n} \rho_i \mathrm{GCF}_{t-1} + \sum_{i=1}^{n} \phi_i \mathrm{TRADE}_{t-1} + \mu_t\end{aligned} \tag{7.2}$$

$$\begin{aligned}\mathrm{GDPG}_t = \beta_0 &+ w_1\mathrm{GDPG}_{t-1} + w_2\mathrm{REM}_{t-1} + w_3\mathrm{POPG}_{t-1} \\ &+ w_4\mathrm{TRADE}_{t-1} + w_5\mathrm{INF}_{t-1} + \sum_{i=1}^{n}\beta_i\Delta\mathrm{GDPG}_{t-1} \\ &+ \sum_{i=1}^{n}\alpha_i\Delta\mathrm{REM}_{t-1} + \sum_{i=1}^{n}\gamma_i\Delta\mathrm{INF}_{t-1} \\ &+ \sum_{i=1}^{n}\lambda_i\Delta\mathrm{POPG}_{t-1} + \sum_{i=1}^{n}\phi_i\Delta\mathrm{TRADE}_{t-1} + \mu_t \end{aligned} \tag{7.3}$$

$$\begin{aligned}\Delta\mathrm{GDPG}_t = \beta_0 &+ \sum_{i=1}^{n}\beta_i\Delta\mathrm{GDPG}_{t-1} + \sum_{i=1}^{n}\alpha_i\Delta\mathrm{REM}_{t-1} \\ &+ \sum_{i=1}^{n}\gamma_i\Delta\mathrm{INF}_{t-1} + \sum_{i=1}^{n}\lambda_i\Delta\mathrm{POPG}_{t-1} \\ &+ \sum_{i=1}^{n}\phi_i\Delta\mathrm{TRADE}_{t-1}\Omega_1\mathrm{ECT}_{t-1} + \mu_t \end{aligned} \tag{7.4}$$

7.5.2.3 Diagnostic Tests

Diagnostic tests are an important element in econometric modelling as they ensure that reliable and credible estimates are derived. To achieve this, the Classical Linear Regression Model (CLRM) assumptions must be satisfied (Pesaran, 1974). This is the only way that the model can provide valid results where inferences can be made. In this research, we conduct various pre-estimation and post-estimation tests. These include testing for stationarity in the different time series variables, checking that the error terms are independent and testing if the variance of the model is constant overtime or homoscedastic. Additionally, the distribution of the data should be normal. In instances where the normality test fails, if the sample size is large enough, the distribution is approximately normal. This arises from the Central Limit Theorem which renders at least 30 observations as a large sample. Lastly, stability tests are also important for the coefficients to ensure that the model does not break down. The stability of the long-run coefficients are tested by the short-run dynamics of the models after the ECM model is estimated.

The ARDL bounds test is based on the assumption that the variables are I(0) or I(1). So, before applying this test, we determine the order of integration of all variables using the unit root tests.

7.5.3 *ARDL Bounds Testing Approach and Analysis*

7.5.3.1 *Stationarity Tests*

Though the ARDL approach to cointegration does not require unit roots pre-testing, determining the maximum order of integration of the variables is necessary because a variable that is stationary at second difference, I(2), cannot fit in the bounds testing (Yakubu et al., 2015). The time series used in the model should either be integrated of order zero I(0), one I(1) or a combination of both. To identify the order in which our variables are integrated, two stationarity tests were employed. First, the Dickey and Fuller (1979) test is conducted and then the Phillips and Perron (1988) test to confirm the results of the Dickey and Fuller (1979).

The results of the Unit root tests are shown in Table 7.2. The results indicate that the variables are a combination of I(0) and I(1) processes. Hence, they can be modelled in an ARDL framework.

Table 7.2 Stationarity test

	Augmented Dickey-Fuller			*Phillips-Perron*		
Variable	*With constant*	*With trend and constant*	*Order of integration*	*With constant*	*With Trend and Constant*	*Order of integration*
GDPG	−3.5925	4.1865	I(0)	−3.5925	−4.1865	I(0)
REM	−3.5966	−4.1923	I(1)	−2.9314	−3.5181	I(0)
CPS	−3.5966	−4.1923	I(1)	−3.5966	−4.1923	I(1)
GXP	−3.5966	−4.1923	I(1)	−3.5966	−4.1923	I(1)
INFLATION	−2.9314	−3.5181	I(0)	−3.5925	−3.5181	I(0)
POPG	−2.6115	−3.5403	I(0)	−2.6049		I(1)
GCF	−3.5966	−4.1923	I(1)	−3.5966	−4.1923	I(1)
TRADE	−3.5925	−4.1865	I(0)	−3.5924	−4.1865	I(0)

Table 7.3 Optimal lag selection

Lag[24]	*LogL*	*LR*	*FPE*	*AIC*	*SC*	*HQ*
0	−508.91	NA	99,302.7100	25.6953	25.9064	25.7716
1	−385.35	210.0499	727.3016	20.7673	22.0340	21.2253
2	−316.14	100.3513	84.7916	18.5569	20.8791	19.3966
3	−242.88	87.9082	8.9459	16.1441	19.5219	17.3654
4	−192.97	47.4208*	3.6366*	14.8983*	19.3316*	16.5012*

The * represent optimal lag structure chosen by the model based on the different criteria

7.5.3.2 *Optimal Lag Selection*

It is important to determine the optimal lag structure for the ARDL model and ensure that the residual of the optimal lag for the cointegrating equation is not serially correlated. There are different information criterions which can be used, depending on if the goal is to arrive at the best model fit or parsimony. Information criterions which are primarily used to select models include the Akaike Information Criterion (AIC), Schwarz Bayesian Criterion (SBC), Hannan-Quinn information among others. Table 7.3 gives the results of the optimal lag section for Eq. 7.4. Based on all the information criterions, a lag length of 4 is the optimal length for Eq. 7.4. This indicates that the residual of the models does not suffer from serial autocorrelation up to lag 4. Therefore, Eq. 7.4 is estimated using a lag length of 7.4.

7.5.3.3 *Bounds Testing for Long-Run Relationship*

The ARDL Bounds test is conducted to determine if there exists a long-run relationship between GDP growth and Remittances received in Jamaica. Results emanating from this test show that the F-Statistic of 8.7214 is greater than the upper bound value for all conventional levels of significance (see Table 7.4). This means that the null hypothesis of no cointegration for the Bounds test can be rejected. Therefore, it can be concluded that there is sufficient evidence to conclude that there exists a long-run relationship between GDP Growth and Remittances in Jamaica.

Table 7.5 shows the long-run coefficients estimated from Eq. 7.4. The model estimates that there is a negative long-run relationship between GDP growth, Remittances, Population growth and inflation at the 1%

[24] Lag Selection Model for Eq. 7.4

Table 7.4 Bound test results

K[25]	*F-Statistics*	*Critical value*	*Lower bound value*	*Upper bound value*
4	8.7214	10	2.45	3.52
		5	2.86	4.01
		2.5	3.25	4.49
		1	3.74	5.06

level of significance. More specifically, a 1% increase in the level of Remittances received by Jamaicans reduces GDP growth by 12.6%. Results emanating from Husien (2019) suggest that remittances have a positive and significant long-run impact on economic growth.

Pesaran and Shin (1996) noted that the estimation of a long-run relationship can be deceptive if the rate at which the model re-adjusts to equilibrium after a shock is not established. The Error Correct Term (ECT) is shown in Table 7.6 and was found to be negative and less than one. These are basic requirements because if the coefficient is greater than one, it may be suggesting that the model is over-specified. Additionally, if the sign on the coefficient is not negative, the model may not revert to its equilibrium after a shock. The coefficient of the ECT in the model is −0.6885 or 69% and is statistically significant at the 1% level. This shows that the speed of adjustment per year is 69%. Therefore, the model adjusts to equilibrium within 2 years or approximately in 17 months after a shock.

Table 7.5 ARDL Long-run estimated coefficient: dependent variable D(GDPG)

Variable[26]	*Coefficient*	*Std. Error*	*t-Statistic*	*Prob*
REM	−1.2596***	0.2846	−4.4265	0.0002
POPG	−17.6225***	4.0661	−4.3341	0.0003
TRADE	−0.0207	0.0688	−0.3007	0.7666
INFLATION	−0.1963***	0.0682	−2.8776	0.0090

The * represent optimal lag structure chosen by the model based on the different criteria

25 ARDL Bounds Test for Eq. 7.4.

26 ARDL Long-run estimated coefficient: Dependent variable D(GDPG).

Table 7.6 Short-run estimates using ARDL approach: dependent variable D(GDPG)

Variable	*Coefficient*	*Std. Error*	*t-Statistic*	*Prob*
C	22.4856	3.2377	6.9449	0.0000
D(GDPG(−1))	0.1350	0.0836	1.6151	0.1212
D(REM)	−0.7362	0.1780	−4.1361	0.0005
D(REM(−1))	1.0285	0.1942	5.2973	0.0000
D(REM(−2))	0.6149	0.1917	3.2085	0.0042
D(POPG)	26.0312	15.5312	1.6761	0.1086
D(POPG(−1))	−152.6918	31.1632	−4.8998	0.0001
D(POPG(−2))	213.7333	28.4391	7.5155	0.0000
D(POPG(−3))	−97.0813	12.7241	−7.6297	0.0000
D(TRADE)	0.0401	0.0243	1.6503	0.1137
D(TRADE(−1))	−0.0331	0.0210	−1.5770	0.1297
D(TRADE(−2))	0.0375	0.0167	2.2410	0.0360
D(TRADE(−3))	0.0545	0.0160	3.3977	0.0027
D(INFLATION)	−0.1846	0.0285	−6.4782	0.0000
CointEq(−1)*	−0.6885	0.0956	−7.2051	0.0000
R-squared	0.9250	Mean dependent var		0.0488
Adjusted R-squared	0.8830	S.D. dependent var		3.2374
S.E. of regression	1.1075	Akaike info criterion		3.3220
Sum squared resid	30.6620	Schwarz criterion		3.9554
Log likelihood	−51.4404	Hannan-Quinn criter		3.5510
F-statistic	22.0198	Durbin-Watson stat		2.0283
Prob(F-statistic)	0.0000			

The * represent optimal lag structure chosen by the model based on the different criteria

In the short run, remittance inflows received reduces output by approximately 7.4% contemporaneously. This may be because remittances are typically used for consumption purposes and remittance receivers may not try to work as hard as they would have to secure an income.

7.5.3.4 *Diagnostic Tests for ARDL Model*

Table 7.7 shows the results emanating from the diagnostic tests on the ARDL model. The Cumulative Sum of Recursive Residuals (CUSUM) and Cumulative Sum of Squares of Recursive Residuals (CUSUMSQ) shown in Fig. 7.6 where panels (a) and (b) show that the coefficients in our model are stable at the 5% level of significance. Panel (c) also gives a pictorial representation of the Jarque-Bera normality test. Based on the

Table 7.7 Diagnostic test

Diagnostic test	*F-Statistics*	*Prob. Chi-Square*	*Prob F*
Breush-Godfrey serial correlation test	0.4173	0.4664	0.7939
Breusch-Pagan-Godfrey heteroscedasicity Test	0.9229	0.4778	0.5646
Jarque-Bera test	0.261	0.8777	
Ramsey RESET test	1.0009		0.3272

probability of 0.8777, we can conclude that the distribution is normal. Overall, all diagnostic tests were passed.

7.5.4 Stability and Normality Distribution Tests

See Fig. 7.6.

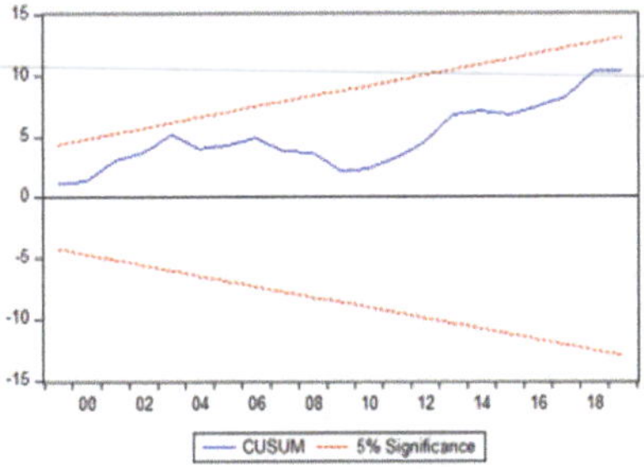

(a) Cusum

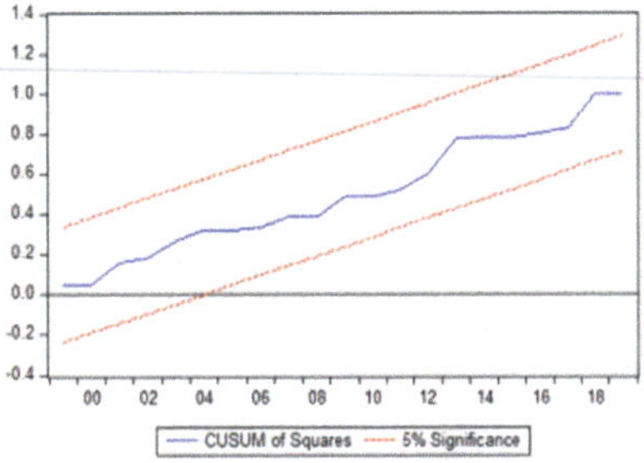

(b) Cusum of squares

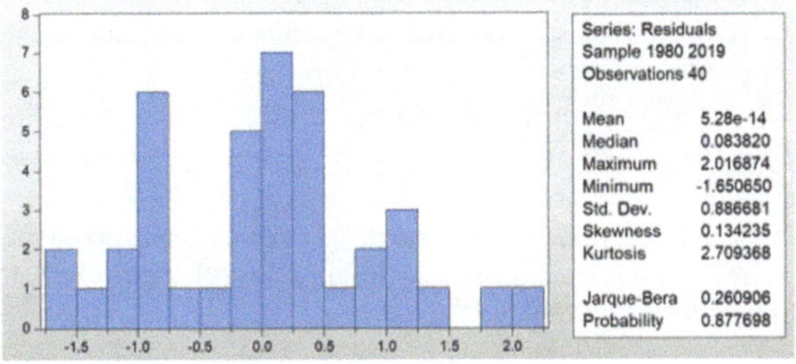

(c) Jarque-Bera Normality Test

Fig. 7.6 Stability and normality test

7.6 Conclusion

Jamaica receives a relatively high level of remittances as a share of GDP in comparison to other countries across the Caribbean. Despite this flow of funds, the ARDL model estimated in this chapter find evidence to suggest that while remittances help to smooth the income of the receiving households, it contributes negatively to the Gross Domestic Product of the country as most of the remittances received are primarily used for consumption purposes. This impact of remittance on GDP is however stronger in the short run which may be an indication of the vast amount of remittances that is consumed. Consumption a component of the calculation of GDP from an expenditure point of view due to Jamaica's high import content, it doesn't seem to contribute to overall GDP growth in Jamaica as indicated by both the long-run and short-run estimates.

References

Adeseye, A. (2021). The effect of migrants remittance on economy growth in Nigeria: An empirical study. *Open Journal of Political Science, 11*(1), 99.

Ajayi, M. A., Ijaiya, M. A., Ijaiya, G. T., Bello, R. A., Ijaiya, M. A., & Adeyemi, S. L. (2009). International remittances and well-being in sub-Saharan Africa. *Journal of Economics and International Finance, 1*(3), 78–84. https://academicjournals.org/journal/JEIF/article-full-text-pdf/F7388961928

Alfieri, A., Havinga, I., & Hvidsten, V. (2006, February). Issue paper: Definition of remittances. In *Meeting of the United Nations technical subgroup on movement of persons–mode* (Vol. 4). https://millenniumindicators.un.org/unsd/tradeserv/TFSITS/TSGMode4/tsg3-feb06/tsg0602-14.pdf

Amuedo-Dorantes, C. (2014). *The good and the bad in remittance flows*. IZA World of Labor. https://wol.iza.org/articles/good-and-bad-in-remittance-flows/long

Anetor, F. O. (2019). Remittance and economic growth nexus in Nigeria: does financial sector development play a critical role? *International Journal of Management, Economics and Social Sciences, 8*(2), 116–135.

Azizi, S. (2021). The impacts of workers' remittances on poverty and inequality in developing countries. *Empirical Economics, 60*(2), 969–991. https://doi.org/10.1007/s00181-019-01764-8

Balance of Payments Manual, Sixth Edition Compilation Guide. https://books.google.com.jm/books?id=q0bnBwAAQBAJ&printsec=frontcover&dq=Balance+of+Payments+Manual,+Sixth+Edition+Compilation+Guide&hl=en&sa=X&redir_esc=y#v=onepage&q=Balance%20of%20Payments%20Manual%2C%20Sixth%20Edition%20Compilation%20Guide&f=false

Cela, T. (2020, May). *Haitian families and loss of remittances during COVID-19 pandemic*. https://oecd-development-matters.org/2020/05/29/haitian-families-and-loss-of-remittances-during-the-covid-19-pandemic/

Chami, R., Fullenkamp, C., & Jahjah, S. (2005). Are immigrant remittance flows a source of capital for development? *IMF Staff Papers, 52*(1), 55–81.

Harris, R., & Sollis, R. (2003). *Applied time series modelling and forecasting*. Wiley.

Hosny, A. (2020). Remittance concentration and volatility: Evidence from 72 developing countries. *International Economic Journal, 34*(4), 553–570.

Husein, J. G. (2019). Foreign aid, workers' remittances and economic growth in Jordan. *International Journal of Social Economics, 46*(4), 532–548.

Imran, K., Devadason, E. S., & Kee Cheok, C. (2020). Foreign remittances and regional poverty: Evidence from household data. *International Migration, 58*(4), 214–230. https://www.researchgate.net/profile/Kashif-Imran4/publication/338009308_Foreign_Remittances_and_Regional_Poverty_Evidence_from_Household_Data/links/5e0c2197a6fdcc28374d4254/Foreign-Remittances-and-Regional-Poverty-Evidence-from-Household-Data.pdf

Ivey, Wendel. (2017). An Empirical Analysis of Minimum Wage on the Quality of Life. *Southwestern Journal of Economics, 12*(1).

Pesaran, M. H. (1974). On the general problem of model selection. *The Review of Economic Studies, 41*(2), 153–171.

Pesaran, H., & Shin, Y. (1999). An ARDL modelling approach to Cointergration analysis'. *Econometrics and economic theory in the 20th century*.

Ramocan, E. G. (2010). *Remittances to Jamaica: Findings from a national survey of remittance recipients*. Bank of Jamaica. https://www.boj.org.jm/uploads/pdf/papers_pamphlets/papers_pamphlets_Remittances_to_Jamaica_-_Findings_from_a_National_Survey_of_Remittance_Recipients.pdf

Roberts, D. (2006). *The development impact of remittances on Caribbean economies*: The case of Guyana. In SALISES Conference, University of the West Indies, St. Augustine. https://sta.uwi.edu/conferences/salises/documents/roberts%20d.pdf

World Bank. (2005). *Global development finance 2005: Mobilizing finance and managing vulnerability, volume 1. Analysis and statistical appendix: global development finance*. World Bank. https://openknowledge.worldbank.org/handle/10986/8135. License: CC BY 3.0 IGO. http://hdl.handle.net/10986/8135

World Bank. (2006). *Global development finance 2006: The development potential of surging capital flows, volume 1. Review, analysis, and outlook: Global development finance*. World Bank. https://openknowledge.worldbank.org/handle/10986/8125. License: CC BY 3.0 IGO. http://hdl.handle.net/10986/8125

World Bank Country and Lending Groups. https://datahelpdesk.worldbank.org/knowledgebase/articles/906519-world-bank-country-and-lending-groups

Yakubu, Y., Salisu, B. M., & Umar, B. (2015). Electricity supply and manufacturing output in Nigeria: Autoregressive distributed lag (ARDL) Bound testing approach. *Journal of Economics and Sustainable Development*, *6*(17), 7–19.

CHAPTER 8

Corruption and the Informal Economy in Jamaica

8.1 Corruption in Jamaica and the Caribbean

In the 2019 Global Corruption perception index, more than two-thirds of the people interviewed in the Caribbean and Latin America believe that corruption has increased over the last 12 months. According to a survey conducted by the Global corruption index, more than 50% of the people interviewed in the Caribbean and Latin America said their government were failing to address the issues surrounding corruption and corrupted officials. The survey results indicated that the police and elected officials were 47% corrupt, the highest on the list. Local governments were listed as 45% corrupt, Prime Ministers/Presidents 43% corrupt, Judges and magistrates 40% corrupt, government officials 37% corrupt, tax officials and business executives 36% and religious leaders 25% corrupt. Jose' Ugaz further highlighted that government must make a concerted effort to root out corruption at all levels to satisfy their citizen's plea for increased accountability and transparency. Jose' Ugaz chair of Transparency International recently declared that the people of the Caribbean and Latin America are being let down by their governments and the private sector. Haughton (2017) also found evidence to suggest that corruption control can improve Jamaica Balance of Payments.

Corruption control measured by the Global Corruption Index (GCI) has been found to have a positive effect on labour productivity in Jamaica with statistical and economic significance based on the results

A. Haughton and W. Ivey, *Overcoming Productivity Challenges in Small Countries*, https://doi.org/10.1007/978-3-031-23301-2_8

of Chapter 2. The GCI measures corruption using data on many factors including briberies, embezzlement, official's use of public position for private gain, official's disclosure of funds etc., for 160 countries including Jamaica.[1] In 2020, Jamaica scored 44 (on a scale of zero most corrupt to one hundred least corrupt) and is ranked the 70th most corrupt country in the world or 5th most corrupt country in the Caribbean. Jamaica's low score on the index is based on the many scandals in the public domain including the gross mismanagement of public funds by government officials. An unaccounted for US$40 million worth of oil at PetroJam and US$1.5 million at the Ministry of Education, as well as widespread nepotism, where friends and relatives are favoured to receive jobs, publicly procured contracts or unwarranted benefits. In general, Latin American and the Caribbean countries are among some of the most corrupted in the world. Like Jamaica, the region has a score of 44 out of 100, no improvement on previous years, which is an indication of a failure to advance towards creating lesser corrupted nations. Transparency International reported that 48% of people interviewed in Jamaica and the Caribbean believed that corruption increased between 2018 and 2020. Barbados appears to be the least corrupt in the region with a score of 68 out of a hundred on the index.

Corruption in Jamaica and around the world is nothing new. Historically, corruption has been evident in medieval societies but became noticeable in the thirteenth century BC, an era which coincided with the reign of the Assyrian civilization.[2] There was no formal definition for the practice at the time, but the act ensued as it was beneficial for the participants to do so. Later, intellectuals like Machiavelli either conceptualized thoughts of or dealt with corruption in some way. In 1621, Sir Francis Bacon was one of the first public officials to be impeached and imprisoned for corrupted behaviour in Britain. The first signs of corruption in Jamaica came while it was a colony of Britain, see Burnard (2004). Settlers

[1] This in addition to including division of public funds, the ability of the government to contain corruption through integrity commission or the strength of the legal system to prosecute them, adequate laws, legal protection for whistle blowers, journalists as well as access to information on public affairs. The index also includes data on meritocratic vs nepocratic appointment of public officials, red tape and excessive bureaucracy that might increase opportunities for corruption by citizens, see Transparency International Global Corruption index methodology.

[2] https://www.britannica.com/place/Mesopotamia-historical-region-Asia/Early-history-of-Assyria.

bribed politicians or their secretaries to get first preferences to acquire land.[3] However, it was not until 1995 that the term corruption was put into perspective after the caveats became apparent to countries and global institutions. The question arises as to what are the causes of this corruption?

8.2 The Causes of Corruption in Jamaica

Dimant and Tosato (2018)[4] found that factors like political monopolistic control, greed, material gains, poverty, low economic and press freedom, political instability, weak property rights, low levels of education, extravagant family and high unemployment are common causes of corruption across countries most of which are characteristics of Jamaica and other SIDs. In a country like Jamaica where income levels are low and cost of living is high relatively, the economic imbalance coerces people to participate in illegal activity including corruption. Šumah (2018) also found evidence to suggest that low income, a less opened economy and non-Protestant religion/faith countries have far more corruption. Jamaica too is a country with high religious beliefs, the moral standards however are very low. There are many scandals of religious leaders who violate their order and commit crimes. The political and economic environment in a country, including the level of professional ethics and morality, habits, customs, traditions and demography can act as enablers for corrupted behaviour (Sumah, 2018).[5] These factors create a breeding ground for citizens to prosper from corruption by increasing the returns with less consequences.

According to Transparency International, the most corrupt countries are characterized by weak democratic systems and are leaning towards a type of authoritarian leadership that uses populism to control the masses while accomplishing their corrupted objectives. Populism enhances support for politicians through publicity stunts rather than effective leadership or policies that promote positive change. These stunts and tactics include:

[3] Burnard, T. (2004). *Mastery, tyranny, and desire: Thomas Thistlewood and his slaves in the Anglo-Jamaican world*. University of North Carolina Press.

[4] https://onlinelibrary.wiley.com/doi/10.1111/joes.12198.

[5] https://www.intechopen.com/chapters/58969.

1. *Undermining the free and independent media, especially when the coverage challenges their message.* Several authors like Dimant and Tosato (2018)[6] and Sumah (2018) found evidence to support TI on this point. However, press freedom is not really an issue in Jamaica, who is ranked 7th out of 180 countries on the world press freedom index.[7] The media is free to report whatever they see fit. The main issue in Jamaica and many other SIDs is that some of the reporters and news networks are corrupt themselves. They accept bribes from politicians to report positive news about them, ignore their negatives and highlight negative news about their competitors. This tends to sway the public's mind to think negative about one group and positive about the other. By influencing public perception, they contribute to swigging the election one way or another.
2. *Use of covert mechanisms to silence and control civil society and international organizations.* This too does not really apply to Jamaica directly. International organizations like the IADB, the World Bank and USAID along with many local civil society groups are very active in Jamaica. Rather, it is the funds from these agencies that are mismanaged and embezzled. Even while they are present overseeing the project, the money is stolen from under their noses. Jamaica has received billions of dollars in Foreign Direct Investments, Official Developmental Aid and loans from global institutions to assist with developmental issues but like Haiti much of this is not mirrored in the economy.
3. *Increase in voter suppression and lessening of certain people's rights to vote.* Voter turnout is a real issue in Jamaica, for the 2020 election only 37% of voters turned out for the election compared to 49% in 2016 a 12% decline. It is however unknown whether the low turnout was because of voter suppression strategies or people deciding to stay home due to the COVID-19 pandemic or people just turned off from the voting process or a combination of all three.
4. *Increase in racist and classist or any other prejudice approach without a concerted effort towards establishing or maintaining equal rights and justice for everyone.* Coming out of slavery, Jamaica has always been a classist country. Even though the system presents itself as

[6] https://onlinelibrary.wiley.com/doi/10.1111/joes.12198.

[7] https://rsf.org/en/ranking.

equitable, people from certain addresses find it difficult to acquire certain jobs based on their dress codes, how they speak and their culture.

5. *An increase in public promises to solve corruption but participation in and continuation of corrupted activities and practices are widespread.* In 2020, the Government of Jamaica were accused of not signing an international agreement to increase the fight against corruption. Considering all the scandals in Jamaica, in recent times none of the politicians arrested are who have been found to breach the laws of the land have faced any severe consequences irrespective of the political party they represent.
6. *Interference in and use of national institutions to weaken the system of checks and balance and increase executive powers.* Of the 63 seats in parliament, the government has 49 while the opposition only has 14, which gives the government a sort of autonomous powers to make certain decisions and pass certain acts. In Mauritius for example, there are checks and balances to mitigate government monopoly power, these will be expounded upon in chapter.

The Caribbean and Latin American countries including Jamaica are characterized by some but not all of these features. In this regard, corrupted officials and business actors sabotage the country's growth and developmental pursuits by abusing their trusted position of power to obtain private benefit.

8.2.1 Corruption and Competition

Corruption can arise from economic agents seeking advantage in markets with less competition during the ordinary functioning of the economy. Ades and Di Tella (1999) examine the causes of corruption across countries by testing the hypothesis that natural rents (i.e. oil) and rents created from lack of competition in the market can foster corruption. In theory, less competition indicates that bureaucrats can extract more rent from the firm they control. The results show that countries with higher rents also have higher corruption. Corruption is especially higher in countries where domestic firms are protected from foreign competition due to barriers to trade, monopoly of a few firms dominating the economy and ineffective antitrust regulations that do not prevent anticompetitive behaviour from the firms. The study claims that the effects are significant, citing big differences in corruption between Italy and Austria, where about one-third of the gap between the countries is due to Italy's lower foreign competition. In Jamaica, PetroJam is the only energy producing company, they are also plagued by one of the biggest scandals. The lack of competition in the energy sector and other sectors in Jamaica might also be factors that contribute to corruption. Ades and Di Tella (1999) conclude that policies that focus on creating more competitive markets can be effective in controlling corruption.

Although the lack of competition can breed corruption in some sense, Bliss and Di Tella (1997) argue that economists have not identified the correct conditions where an increase in competition reduces corruption. This is so because competition is not an exogenous parameter that can be easily varied in a model to determine how corruption levels change due to simultaneous effect between corruption and competition. Bliss and Di Tella (1997) empirical analysis does not lead to a concrete answer about whether an increase in competition reduces corruption or not. The study explains that the ambiguity of the results is due to the "uncertainty about the costs that the corrupted officials face". This cost can range from zero if they are not caught and prosecuted to millions of dollars if they are. The former is normally the case for the majority. Only a minority of politicians and public officials are arrested in Jamaica, and very rarely are they prosecuted. More recently, Jamaica arrested and charged its Minister of Education for embezzlement. The case is still ongoing. Despite the ambiguity in their results, Bliss and Di Tella (1997) argued that the adverse effects of corruption and the abundance of corruption agents within countries can be reduced by increasing competition in product markets.[8]

8.3 Corruption Greasing the Wheels

There has been a solid debate in the economic literature as to whether corruption is good or bad for the economy. A large portion of the corruption literature analyses the impediments of corruption, but some cite that corruption saves time and reduces bureaucracy in particular instances.

[8] Not necessarily corruption on a whole, the level of corruption can be the same but the adverse effects lessened...

Marie (2014) outlined that while there is a large consensus in the literature on the negative impact of corruption on economic growth, some researchers continue to argue that the effect of corruption on growth is context-specific. In some highly regulated countries that do not have effective government institutions and governance systems, corruption can compensate for red tape and institutional weaknesses and "grease the wheels" of the economy. Many theories posit that corruption can act as the lubricant of the economic wheel and at least in some cases have a positive impact on economic growth. Leff (1964), Huntington (1968) and Lui (1985) argue that corruption increases efficiency via two channels. It speeds up the governmental process and introduces competition to scarce government resources. Leff (1964), Huntington (1968) and Lui (1985) postulated that a system built on bribery for allocating these scarce resources leads to an outcome in which the most efficient firms can pay the highest bribes. This might be a rationale why corruption acts as a functional part of the doing business environment in Jamaica and many other Caribbean and Latin American economies. It can be found on all levels of the household, within firms and across the government.

In low-income countries, some people become innovative to survive, much of this innovation comes in the form of corruption. In some instances, employees create additional bureaucracy in many processes to slow down the transactions deliberately to attract bribes. In other instances, bribes are automatic means to speed up any process. According to Myrdal (1969), corrupt bureaucrats/politicians, instead of speeding up the governmental process, slow the process down to create more opportunities for bribery, see also, Rosa et al. (2015) and Teoman et al. (2021). Bribes are like gratuities that motivate people, government and businesses to get things done. You must "grease the wheel for it to spin" is a common analogy used. Almost every business process that requires waiting time presents an opportunity for bribery or corrupt behaviour. For example, the process of clearing goods from the wharf in Jamaica is sometimes very bureaucratic and cumbersome depending on who you are and who you know. In this case, people receive their goods cheaper with less lead time by bribing the employees, who now receive a higher pay for doing their job more efficiently. In Jamaica, depending on the circumstances, corruption plays both an enabling and disabling role (though the net effect is yet to be empirically determined) in economic activities and productivity performance.

Corruption is also present in other situations like getting a driver's licence, renewing motor vehicle licence and fitness or even to avoid traffic tickets when stopped by the police. In Jamaica for example, people rarely complete the driver's licence test, many people bribe the examiner to pass or buy their licence as it is called. There is a classic story about a professional driver who drove without a licence for 20 years, each time he did the test he failed. After failing his driving test for the 5th time, he inquired the reason why he failed so many times even after following the instructions correctly. He was told he didn't pay the bribe. He paid the examination fee but didn't pay the examiner in the car beside him a fee. This is the case for many people in many places in these countries and there are many other examples. Heavy bureaucracy in government and other processes create loopholes for workers to earn additional income to speed up the process. A more subtle form of corruption is found in congested establishments where more recent entrants or people at the back of the queues apart from pregnant ladies and the elderly are allowed to skip the line simply because they have some form of connection with people who work in the establishment. Despite the presume benefits of corruption, there are consequences. For example, one of the leading unnatural causes of death in Jamaica is road fatalities which claim over 300 lives on average each year. While there is no empirical evidence to show that this is direct linked between perpetrators who cannot drive properly because they didn't learn how to and simply purchased their licence, it is likely that drivers who pass the test and follow the procedures are less inclined to have accidents. But the consequences of corruption are more far reaching than deaths on the road, corruption can kill a country's economic developmental objectives.

8.4 The Consequences of Corruption

Myrdal (1969) found evidence suggesting that corruption can be a major hindrance to a country's developmental pursuits. Tanzi and Devoodi (2002) found negative relationship between corruption and public revenues. Their results also showed that corruption leads to substandard infrastructure, like inferior quality roads and higher electricity outages. Corruption reduces the effectiveness of public investments and the infrastructure of a country. Olson et al. (2000) argue that differences in governance can explain why most low-income countries do not grow faster than the high-income countries. The results show that countries with better governance have higher productivity growth. Olson et al. (2000) reiterate that a government of integrity main aim should be to allocate the economy's scarce resources efficiently rather than indulging in rent-seeking and corrupt activities that only benefit both parties in the short run from a microeconomic standpoint. This has serious implications for productivity. Consequently, the rent-seeking activities would hinder developing countries like Jamaica from achieving their sustainable growth and development goals.

Corruption can slow down the process of national development by reducing productivity as shown by the result of Chapter 2 in the case of Jamaica. Teoman et al. (2021) also found evidence to suggest that corruption decreases productivity in the long run in Turkey. Using data from 1950 to 2014, their results show that the adverse effects worsened with the size of the government. Intuitively, larger governments imply an increase in the number of corrupted parliamentarians/politicians who are seeking rents from illegal behaviour, distorting the revenue earned from taxpayers that could have been reinvested into the macroeconomy. Rosa et al. (2015) highlights that the business environment plays a role in determining the impact of firm-level corruption on productivity. Their results show that in highly corrupt environments where the legal framework is weak, firms that do not pay bribes are more productive. The negative effect of corruption also increases in organizations where the level of institutional quality is declining. The main disadvantage or gap of their analysis is that the results are only available for a cross section in 2009 only, therefore no inference can be gathered from evidence on firm performance over time or the long-term effects.

Bribes are a popular form of corruption in Jamaica. Rosa et al. (2015)'s results show that corruption characterized as "Bribe Tax" has a negative impact on firms' performance. Dismissing the ideology that corruption is a necessary evil that involves "greasing the wheels of commerce" to mitigate the negative effects of inefficient bureaucracy known as "Time Tax". In any economy, corruption coined as "Bribe Tax" increases the firm's opportunity cost; money spent on bribery could have been invested in a productive and efficient manner elsewhere and gain significant returns. This increase in opportunity cost has a negative impact on firms' ability to generate normal profits (perfectly competitive markets) and economic profit (imperfect markets: monopolistic, oligopoly and monopoly) in turn hampering productivity and industries' contribution to national income. According to Shleifer and Vishny (1993), corruption in the form of bribes inflates the cost of goods and services and leads to welfare losses in demand. Corruption acts like a tax that deters business investment and undermines bureaucracy or trust in government institutions. In Jamaica, extortion is prevalent in some towns where businesses pay Care and Protection Tax (CPT) to gangsters or risk theft and loss of life in their business. In this environment, entrepreneurs might be afraid to start businesses or might be forced to shut down or relocate.

Lambsdorff (2003) developed a theoretical model to investigate the relation between corruption and labour supply. The model shows that corruption affects labour supply in the formal sector by reducing productivity and reorienting the supply of labour to the shadow economy, altering the tax burden and distorting the saving-consumption trade-off. The predictions of the theoretical model are tested using panel data methods for 132 countries. Their empirical results show that corruption has a statistically significant robust direct negative effect on the Labour Force Participation Rate (LFPR)[9] and the employment to population ratio. Corruption also has an indirect effect on the LFPR and employment to population ratio through a higher tax burden and increase in size of the shadow economy. Higher wages, an increase in consumption and better regulatory quality are found to reduce the negative impact of corruption on labour supply. However, the overall effect on labour supply is negative, suggesting that the negative effects of corruption outweigh the positive effects of improved regulatory quality, wages and higher consumption. Lambsdorff's (2003) findings imply that to reduce the negative effect of corruption on labour supply, governments need to develop a comprehensive approach, not only to combat corruption itself but also to improve the regulatory framework and to promote policies that decrease activities in the shadow economy.

Hall and Jones (1999) reasoned that, differences in productivity correspond to differences in social infrastructure across countries. Here, social infrastructure are the institutions and government policies that determine the incentives for individuals and firms in each economy. These incentives can result in productive activities (e.g. accumulation of skills, development of new production techniques) or in predatory activities (e.g. rent-seeking and corruption). Hall and Jones's (1999) empirical evidence from 127 countries shows that social infrastructure enables countries to invest in

[9] Used as a proxy for labour supply in the analysis.

human and physical capital at a higher rate and to use these inputs more productively to produce high levels of output per worker. In general, countries with policies that favor productive activities produce more output per worker. For example, the findings suggest that the observed difference in the social infrastructure between Niger and the United States is enough to explain the 35-fold difference in output per worker. Overall, the empirical evidence suggests that countries with weak social infrastructure, corrupt government officials, severe barriers to trade, poor contract enforcement and government interference in production will not reach the levels of output per worker such as in countries in western Europe, northern America and eastern Asia.

8.5 Jamaica's Inaction

Jamaica has established several institutions over the last forty years to control the level of corruption in the country but many of their attempts have been futile. These institutions include the Parliamentary Integrity of Members Commission (1975) to police members of parliament assets and liabilities; the Contractor General (1983) to ensure that the government contracts are awarded properly and the Commission for the Prevention of Corruption (2001) to keep check on the public servants' assets and liabilities. However, each of these institutions has expressed concern regarding the inadequacy of their impact (NIA, 2013). For example, the Commission for the Prevention of Corruption in its 2012–2013 report expressed that there were concerns regarding the backlog in relations to the rulings of suspected "illicit enrichment". NIA (2013) added that there were cases where the Contractor General had requested the prosecution of some public officials, but this has been a challenge. The political regimes have been rapidly losing their legitimacy and this has been undermined by their long history of corruption in the election where both the economic and political elite are implicated in the misuse of funds and other types of corruption (Harriott, 2001). In an atmosphere in which freedom of the press is notable, there is evidence that suggests that this tendency will continue (Blackburn and Powell., 2011). Challenges like these are among the reasons why Jamaica has not achieved much economic growth as Mo (2001) posited that a 1% increase in the corruption level reduces the economic growth rate by about 0.72%. Data from the World Bank indicates that Jamaica has not been doing enough to challenge corruption. Over a 21-year period (1996 and 2019), Jamaica only has positive score for only 3 years. In essence, 86% of corruption scores over the last 2 decades are negative which provides evidence to suggest that Jamaica has a lot of work to do as a nation to reduce or eradicate corruption. The effect of corruption is far reaching and has many dimensions related to political, economic, social and environmental effects (NIA, 2016). Therefore, the level of corruption that exists in a nation can determine whether the country prosper or suffer from economic retardation. Mo (2001) added that corruption primarily affects economic growth through political instability. Therefore, with Jamaica being politically unstable as the data suggest in, coupled with high levels of corruption, it is no accident that the economy is growing at such a low rate.

Efforts are still being made to tackle corruption via the National Integrity Action (NIA). The NIA advocates for the passage and enforcement of these and other laws: Single Anti-Corruption Agency (Integrity Commission), Revision of the Political Code of Conduct, Political Party Financing and Registration, Award of Government Contracts, Public Education, Capacity Building among others. However, the NIA does not have the authority to prosecute. There have been cases where senior public officials and a few private sector individuals were brought before the courts, but successful prosecution remains the exception, with only one recent conviction of note (NIA, 2016). Jamaica has also formed the Major Organised Crime and Anti-Corruption Agency (MOCA) in 2014 which brought together the Anti-Corruption Branch (ACB) and the Major Organised Crime and Anti-Corruption Task Force (MOCA TF). This is an elite agency with a focus on tackling corruption in the public sector and bringing high-valued targets to justice (MOCA, 2016). Data from the World Bank suggests that the rule of law in Jamacia is not respected Fig. 8.1. If the rule of law is violated and corrupted officials do not suffer the consequences, then the institutions established would have been a failure and the country will be in a repetitive cycle characterized by political instability, high levels of corruption and low economic growth which will ultimately be translated into low living standards over time for Jamaicans. Shleifer (1993) found evidence to suggest that corruption increases participation in the informal economy.

8.6 The Informal Economy in Jamaica

Many businesses in Jamaica are unregistered and do not have the necessary paper works to regularize themselves and therefore classified as informal. The informal economy is described in many ways by different

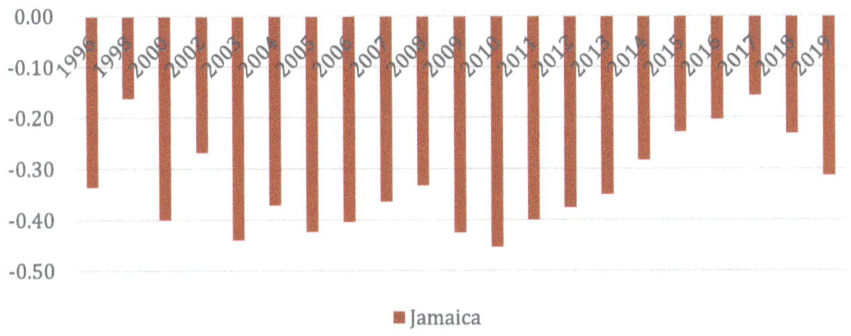

Fig. 8.1 Rule of law (1996–2019) (*Source* World Bank)

people as black, shadow, hidden, irregular, unofficial, unobserved, underground and grey economy (Khuong et al., 2020). Khuong et al. (2020) note that the informal economy encompasses those activities that necessitate a cost but are excluded from the benefits and rights of formal deeds. The International Labour Organization (ILO) sets the criteria for which the Statistical institutes or Labour departments within the government of countries should follow when developing household surveys to derive labour market and decent work statistics and indicators. This also includes how the informal economy is measured. Based on the Seventeenth International Conference of Labour Statisticians (ICLS), employees are considered to have informal jobs if their employment relationship is not subject in law or in practice to: national labour legislation, income taxation, social protection, entitlement to certain benefits (paid annual leave, paid sick leave, maternity or paternity leave, severance pay, etc.). Figure 8.2 shows a flowchart of how the ILO classifies the formal and informal sectors. Although informal, these businesses provide meaningful employment and income and is accounted for in the calculation of Jamaica's GDP.

The ILO estimated that over two (2) billion people, the equivalent of sixty-one per cent (61%) of the world's employed labour force makes their living in the informal economy. A significant proportion (93%) of the globe's informal employment comes from developing and emerging economies (ILO, 2018). High incidence of informality in all its forms has multiple adverse consequences for workers, enterprises and societies and is a major challenge for the realization of decent work for all and to facilitate sustainable and inclusive development (ILO, 2018). It would be beneficial

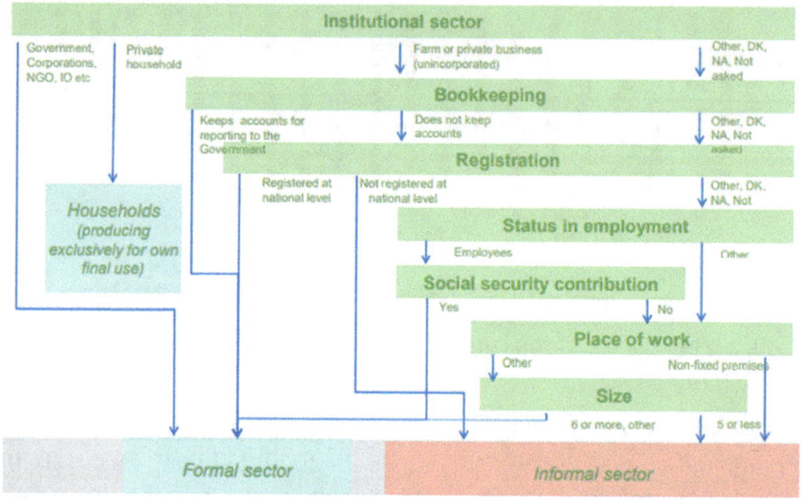

Fig. 8.2 Economic flow chart of the formal and informal sector (*Source* ILO)

to Jamaica and other SIDs to transition their informal economies to a formal one but this is an arduous task.

In Jamaica, the Statistical Institute of Jamaica (STATIN) provides estimates of the size and distribution of informal employment in Jamaica which refers to persons whose main job lacks certain protection and benefits. Officially, Fig. 8.3 shows that forty six per cent (46%) of the labour force are employed to the formal economy, thirty nine per cent (39%) to the informal economy and 15% works in agriculture. Jamaica's informal economy is 7% larger than the formal economy both less of agriculture. STATIN does not dichotomies agriculture into its formal and informal components. Agriculture operates in the rural area, based on eye estimate. For accountability, this is an issue that should be addressed. Information should be present on what percentage of the Jamaican agricultural sector is formal vs what is informal. However, based on the factors that contribute to a business being informal, a large percentage of Jamaica's Agricultural Sector fit those descriptions. Considering that 46% of the economy is firmly categories as being formal, then the remaining Unofficially, the informal economy in Jamaica is much larger figure if the Agricultural Industry is included. For example, if we split agriculture in

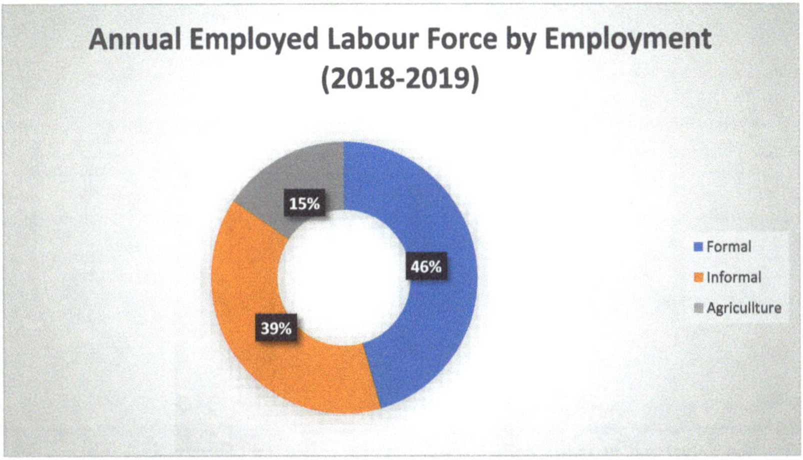

Fig. 8.3 Employed labour force by types of employment (*Source* STATIN)

half formal and half informal, then Jamaica's informal economy will be 53% and its formal is 47%.

STATIN was able to dissect all the other industries in their formal and informal components except agriculture. The Wholesale and Retail Trade, Repairs; Installation of Machinery and Equipment (WRTRM) industries are more informal than formal and contributes approximately 17% of total value-added GDP; the highest contributor to total value added in the Jamaican economy. The industry also employs 20% of the total labour force. Therefore, is not surprising that the industry is also the primary employer of informal labour. Fig. 8.4 shows that the WRTRM industry employs twice as much informal relative to formal labour. The industry is estimated to account for 15% of total formal and approximately 33% of the total informal labour supply. If it formally contributes to 17% of GDP, then it might contribute to 34% of the informal economy based on employment numbers. Construction and Private households with employment have the highest informal to formal employment ratio. The informal construction industry is more than 5 times that of the formal. Public Administration and Defence, as well as Education and Real estate and Lending are some of the least informal industries in Jamaica. Their operations are mostly formal.

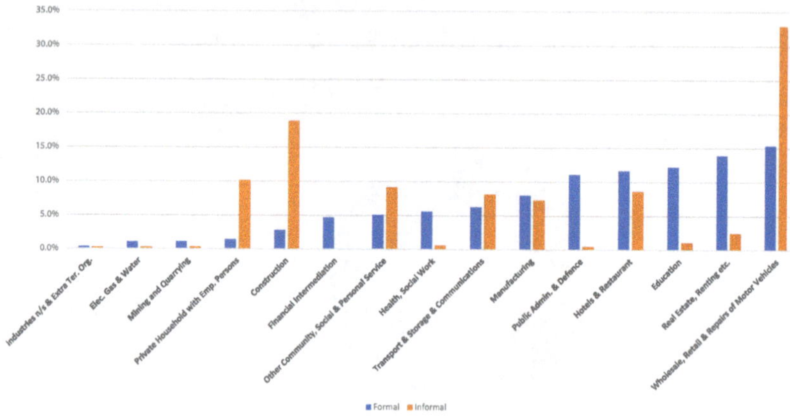

Fig. 8.4 Formal vs informal employment by industry (*Source* STATIN)

8.7 Discussion and Recommendations

Jamaica like every other small country is endowed with resources, including land and labour. These factor supplies by themselves have intrinsic value but must be unearthed properly to generate wealth for the country. Coordinated economic activity between the firms, the government and the household, according to Keynes (1937) can advance any country's pursuits by establishing a consensus that there is something collective to work towards. However, Lucas and Sargent, (1981) points out that rational economic agents develop expectations in their individual pursuits that are sometimes incongruent with nationally oriented policies and intentions. These individuals' rational decisions are innately selfish by nature and can obscure the selfless patriotism required to achieve holistic national development. This individual selfishness called greed, according to Stiglitz (2009) is also the source of some economic downturns. However, with the right consensus and understanding of the process of economics, people might be willing to compromise, if they are able to understand that the whole is greater than the sum of its parts thereby realizing that there are more economic benefits to be had from pursuing national developmental objectives above individual selfish gains from corruption. The latter will always exist, but a collective push towards holistic advancement from the former can improve the choice basket (standard of living) of every economic agent and thereby achieve the required level of economic development. Thus, should be desired and pursued. There are many literatures that highlight the best ways to reduce corruption in a country,[10] see for example, the IMF's International Cooperation Against Corruption which highlight that leader are willing in principle but not by action to reduce corruption. These leaders must take action to implement the necessary strategies with conviction. The overall findings from the different literatures suggest that Caribbean countries specifically Jamaica needs to implement stringent and efficient rules of law and regulations with strengthen institutional organizations as well as encourage competition among the different industries within the economy to mitigate corruption in turn enhancing the productivity capabilities of firms and aggregating the economy.

The progression of a country depends on the strength of its institutions. Less corruption sends a better signal to the rest of the world about the state of leadership and integrity of a country. Decisive actions need to be taken to reduce corruption, mitigate any threats to the democracy of the nation and transition actors in the informal economy to the formal. Transparency International are encouraging citizens, leaders and civil society advocates to continue their efforts by highlighting the gains of low corruption, and advocate for more checks and balances in government-owned entities. Citizens as well should continue to look out for political speeches that falsely use the fight against corruption to continue its publicity stunts while continuing to undermine democracy and increase their authoritarian or individualistic powers. In this modern world, blockchain technology can be used to increase check and balances and reduce the level of corruption in areas where it is most pervasive like in the procurement of government contracts.

References

Alberto, A., & Di Tella, R. (1999). Rents, competition, and corruption. *American Economic Review, 89*(4), 982–993.

Bliss, C., & Di Tella, R. (1997). Does corruption kill corruption? *The Journal of Political Economy, 105*(5), 1001–1023.

[10] https://www.imf.org/external/pubs/ft/fandd/1998/03/pdf/klitgaar.pdf, https://www.oecd.org/corruption/putting-an-end-to-corruption.pdf, https://www.econstor.eu/bitstream/10419/79084/1/729292614.pdf.

Blackburn, K., & Powell, J. (2011). Corruption, inflation and growth. *Economics Letters, 113*(3), 225–227.

Dimant, E., & Tosato, G. (2018). Causes and effects of corruption: What has past decade's empirical research taught us? A survey. *Journal of Economic Surveys, 32*(2), 335–356.

Hall, R. E., & Jones, C. I. (1999). Why do some countries produce so much more output per worker than others? *Quarterly Journal of Economics, 114*(1), 83–116.

Harriott, A. (2001). The crisis of public safety in Jamaica and the prospects for change. *Souls, 3*(4), 56–65.

Haughton, A. (2017). *Developing sustainable balance of payments in small countries: Lessons from macroeconomic deadlock in Jamaica*. Springer.

Huntington, S. (1968). *Political order in changing societies*. Yale University Press.

Keynes, J. M. (1937). The general theory of employment. *The quarterly journal of economics, 51*(2), 209–223.

Khuong, N. V., Shabbir, M. S., Sial, M. S., & Khanh, T. H. T. (2020). Does informal economy impede economic growth? Evidence from an emerging economy.

Lambsdorff, J. G. (2003). How corruption affects productivity. *Kyklos, 56*(4), 457–474.

Leff, N. (1964). *Economic development through bureaucratic corruption*. American Behavioral Scientist.

Lucas, R. E., & Sargent, T. J. (Eds.). (1981). *Rational expectations and econometric practice* (Vol. 2). U of Minnesota Press.

Lui, F. (1985). An equilibrium queuing model of bribery. *Journal of Political Economy, 93*, 760–781.

Marie, C. (2014). The impact of corruption on growth and inequality.

MOCA. (2016, nov). Major organised crime and anti-corruption agency: About us. Retrieved from http://www.moca.gov.jm/

Myrdal, G. K. (1969). The Soft States of South Asia: The Civil Servant Problem. *Bulletin of the Atomic Scientists, 25*(4), 7–10.

NIA. (2013). National integrity action: Why jamaica urgently needs an effective integrity commission. Retrieved from https://niajamaica.org/issues/campaign-financing/22-issues/901-why-jamaica-urgently-needs-an-effectiveintegrity-commission

Olson, M., Sarna, N., & Swamy, A. V. (2000). Governance and growth: A simple hypothesis explaining cross-country differences in productivity growth. *Public Choice, 102*(3/4), 341–364.

Pesaran, M. H., Shin, Y., & Smith, R. (2001). Bounds testing approaches to the analysis of level relationships. *Journal of Applied Econometrics, 16*, 289–326.

Rosa, D. D., Gooroochurn, N., & Görg, H. (2010, August 30). *Is it worth it: Does corruption influence firm productivity?* VoxEU.org.

Rosa, D. D., Gooroochurn, N., & Görg, H. (2015, April 1). Corruption and productivity: Firm-level evidence. *Jahrbücher Für Nationalökonomie Und Statistik, 235*(2), 115–138.

Shleifer, A., & Vishny, R. (1993). Corruption. *Quarterly Journal of Economics, 108*(3), 599–617.

Stiglitz, J. E. (2009). The current economic crisis and lessons for economic theory. *Eastern Economic Journal, 35*, 281–296.

Šumah, Š. (2018). Corruption, causes and consequences. In *Trade and Global Market*. IntechOpen.

Tanzi, V., & Davoodi, H. R. (2002). Corruption, public investment, and growth. In G. T. Abed & S. Gupta (Eds.), *Governance, corruption, & economic performance* (pp. 280–299). International Monetary Fund.

Teoman, O., Dincer, O. C., & Cavusoglu, T. (2021, January 14). *Corruption and total factor productivity in the long run: Evidence from post-WW2 Turkey*. SSRN 3726989.

CHAPTER 9

A Growth Comparison of Mauritius and Jamaica

9.1 Introduction

Many countries around the world are challenged by different problems as it relates to the advancement of their total population. These may include low rates of economic growth (Jamaica), high levels of poverty (Suriname), high levels of income inequality (Japan) and in some cases highly corrupt government officials (Afghanistan). Over the last 30 years, real per-capita GDP in Jamaica increased at an average rate of just 1% per year, making Jamaica one of the slowest growing developing countries in the world (World Bank, 2016b). Other countries have been able to overcome their innate challenges to achieve economic growth, transformation and development (China, Japan, Mauritius, Taiwan, etc.). One such country, Mauritius, a small island in the Oceanic Basin, has grown by 4.4% between 1990 and 2019, 340% or almost 3 ½ times more than Jamaica. The objective of this chapter is to identify aspects of Mauritius's successful transformational developmental strategy that has facilitated their economic growth that Jamaica can learn from.

9.1.1 *Why Do Economies Need to Grow?*

Why is it imperative to target economic growth and not poverty, income inequality or even unemployment? Historically, nothing has worked better than economic growth in enabling societies to improve the life chances

A. Haughton and W. Ivey, *Overcoming Productivity Challenges in Small Countries*, https://doi.org/10.1007/978-3-031-23301-2_9

of their members, including those at the very bottom (Rodrik, 2008). Adams (2003) used data from 50 developing countries and found that economic growth represented an important factor in reducing poverty in the developed world. China alone has lifted over 450 million people out of poverty since 1979 which can be attributed to rapid economic growth between 1985 and 2001 (Lin, 2003). Economic growth can also be useful in reducing the disparity of income inequality between the rich and the poor. For example, it has been estimated that a one per cent increase in income levels could result in a 4.3% decline in poverty in countries with very low inequality or as little as a 0.6% decline in poverty in highly unequal countries (Ravallion, 2005).

Economic growth also has other benefits, such as improved health and education. Bhalotra et al. (2006) found that in India, a 10% increase in GDP is associated with a reduction in infant mortality of between 5 and 7% while HM Treasury (2005) posited that primary and secondary school enrolment rates are positively associated with higher levels of per capita income. It is quite evident that economic growth is integral in advancing the livelihoods of citizens of the world. However, in achieving the best results for a nation, there is no panacea as countries are of different characteristics (governments, initial endowments, geography, demographics, etc.) and have different relationships with each other. This chapter will zoom in on Jamaica and Mauritius which share certain similarities but are growing at different rates. Mauritius has averaged 4% annually in comparison to Jamaica's 1% between 1977 and 2019.

In July 2020, Mauritius was reclassified as a high-income country by the World Bank as the nation achieved the minimum level of GNI per capita for high-income nations. According to the World Bank, the nation reached this milestone in one of the worst years in its history due to the global COVID-19 (coronavirus) pandemic, which has wreaked havoc on its economy.[1] While their progress is likely to be negatively impacted by the COVID-19 pandemic which is dubbed as one of the worst pandemic in history, Mauritius is likely to continue their development after the crisis has dissipated.

[1] World Bank (March, 2021) Mauritius Overview https://www.worldbank.org/en/country/mauritius/overview#1.

9.1.2 *Why Mauritius and Jamaica*

Mauritius has similar endogenous characteristics to Jamaica. Both countries are small economic states which severed ties with British colonial power in August 1962 and March 1968, respectively. Both countries are members of the United Nations (UN) and have a similar parliamentary democratic style of government structure. Historically, both countries' economies reflected a plantation economy and sugar cane was among the chief crops cultivated. In the seventeenth and eighteenth centuries, Jamaica became the major producer and leading exporter of sugar in the world and recorded an all-time high of 514,825 tonnes in 1965 (Sugar Industry Authority of Jamaica, 2000). On the other hand, Mauritius made constant efforts to improve and maintain the economic viability of the sugar cane industry which was virtually the lifeblood of the economy where production reached 150,000 tonnes in 1862 from 52,000 hectares of cane (GOVMU, 2005). In more contemporary times, both economies have been relying heavily on services. Based on Nordea (2021) and UNDP (2020) estimates, services accounted for 67.7% and 72.6 per cent of Gross Domestic Product (GDP) for the Mauritian and Jamaican economy, respectively.

In Mauritius, economic growth was intensive between 1982 and 1990 induced by an increase in labour and capital which together accounted for 90% of the annual 6.2% growth of GDP (Subramanian & Roy, 2001). Additionally, in the 1990s, economic growth was driven by increases in Total Factor Productivity (TFP). TFP, which captures the overall efficiency of production, grew by 1.4% on average annually which accounted for a full 25% of economic growth for the period (Subramanian & Roy, 2001). In contrast, Jamaica's TFP as explained in Chapters 1 and 2 and also in Blavy (2006) has been on the decline since the early 1970s to the mid-1980s and then throughout the 1990s which he believes is a plausible explanation for the sluggish performance of the Jamaican economy over the last few decades. TFP has declined at an average rate of 0.3% annually over the period 1990 to 2019. Similarly, labour productivity, which measures output per worker, contracted at an average rate of 0.33% annually over the same period. Several factors have contributed to Jamaica's low productivity over the years. These include deficiencies in human capital, high levels of crime, fiscal distortions and poor work ethic (Wynter, 2011). It will be interesting to see if these factors have significantly caused the divergence in the TFP growth and in extension

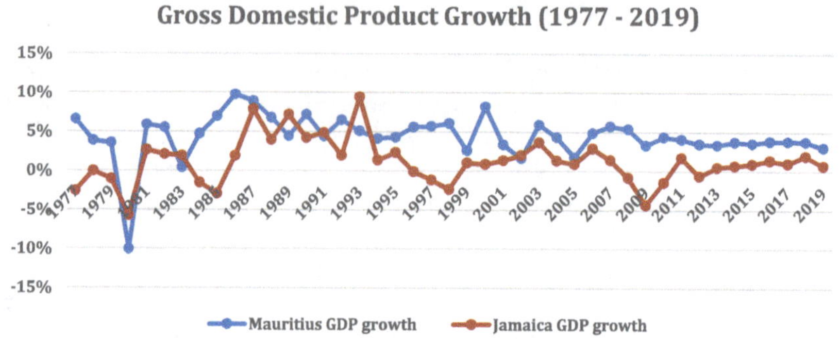

Fig. 9.1 GDP growth in Mauritius and Jamaica (1977–2019) (*Source* World Bank)

the difference in economic growth achieved by the two nations when the cross-country contextual analysis is done in the following section.

Based on the data coming from the World Bank's World Development Indicators (WDI), Fig. 9.1 gives a graphical representation of the GDP growth rates for Mauritius and Jamaica from 1977 to 2019.

The comparison between the two countries can be made from the late 1970s, but the data prior to the 1970s is still relevant in understanding where Jamaica was coming from in contrast to where they are today relative to Mauritius. On average, Mauritius has been growing at a higher rate than Jamaica especially from 1995 onwards. However, having higher growth rates does not necessarily translate to being better off economically. However, Fig. 9.2 shows that there has been a divergence in GDP per capita between Jamaica and Mauritius. Both countries tracked each other closely prior to 1996 where Jamaica had a slight edge over Mauritius. After 1996, Mauritius overtook Jamaica and the gap started to get progressively wider between the two nations. The following sections will discuss some of the factors that may have contributed to this diverge in income per capita.

The additional sections of the chapter are organised so that Sect. 9.2 gives a contextual analysis on the basis of institutions, culture and ethnicity and other special factors that have propelled Mauritius forward relative to Jamaica. Consequently, conclusions are made in Sect. 9.4 and the necessary policy recommendations are made.

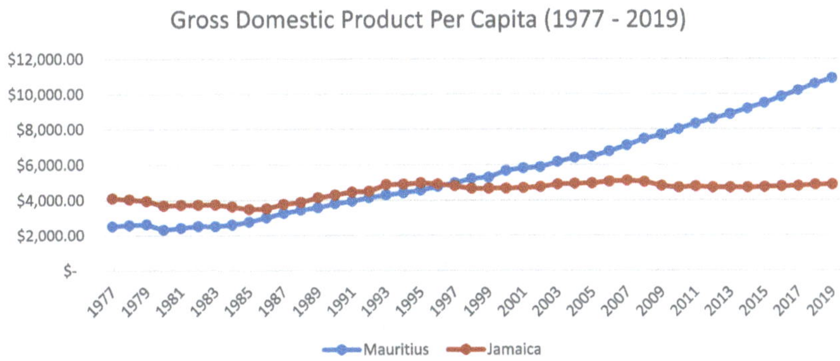

Fig. 9.2 GDP per capita in Mauritius and Jamaica (1977–2019) (*Source* World Bank)

9.2 Cross Country Contextual Analysis

9.2.1 Institutions and Corruption

North (1990) described institutions as the humanly devised constraints that shape human interactions which structure incentives in human exchange, whether politically, socially or economically. Economic institutions are integral to economic growth because they influence investments in physical and human capital, technology and the organization of production (Acemoglu et al., 2005). The (IMF, 2003) in its World Economic Outlook report used data from 1960–1998 in a regression model to find the relationship between economic growth, institutions and policies. The results indicated that institutions play a dominant role in explaining cross-country differences in growth, although macroeconomic policies are also significant. Therefore, institutions could be a factor that gives some insight into why some countries are struggling economically while others are prospering. For example, greater equality and functional economic institutions are seen as the cause for the successful development of Vietnam compared to a similar country as Nicaragua, characterized by high inequality where power is concentrated in the hands of a limited elite, and governments have failed to adequately invest in infrastructure and public welfare (Ferrini, 2012). Similarly, institutional capacity to exploit domestic primary resources is indicated as the key to the success of Botswana in comparison to other developing countries for which primary resources have not proven to be beneficial. These countries include Sierra

Leone (diamonds), Angola, Equatorial Guinea and Nigeria (oil) (Birdsall et al., 2005).

The World Bank's Global Competitive Index for 2019 ranks Mauritius at number 29 while Jamaica is ranked at number 91 out of a total of 141 countries globally in terms of institutional quality (WEF, 2019). Mauritius has an institutional quality score of 64.7, while Jamaica has a score of 49.3 where a score of 100 is the highest on the scale used to measure the quality of institutions. Subramanian and Roy (2001) and Yao et al. (2005) highlighted the high quality of the Mauritian institutions in their work. On the other hand, Duncan-Waite and Woolcock (2008) argued that Jamaica is neither cohesive nor democratic and has displayed context-specific social division and institutional weakness. They noted that the Transparency International (TI), which is the global anti-corruption watchdog, revealed that Jamaica has never scored higher than 3.3 on its Corruption Perception Index, where a score of 10 is perceived to be "clean", and 1 as "corrupt". Figure 9.3 shows the difference in control of corruption for 1996–2019 between Jamaica and Mauritius.

The data may be indicating institutional weaknesses in Jamaica as expressed by Duncan-Waite, and Woolcock (2008) has contributed to the political instability as seen in Fig. 9.4. The inability of Jamaica to adequately control corruption as shown in Fig. 9.3 might be a plausible reason for the disparity in economic growth as the data suggest that Mauritius has been more politically stable and has more control over corruption. Additionally, they have been performing better than Jamaica

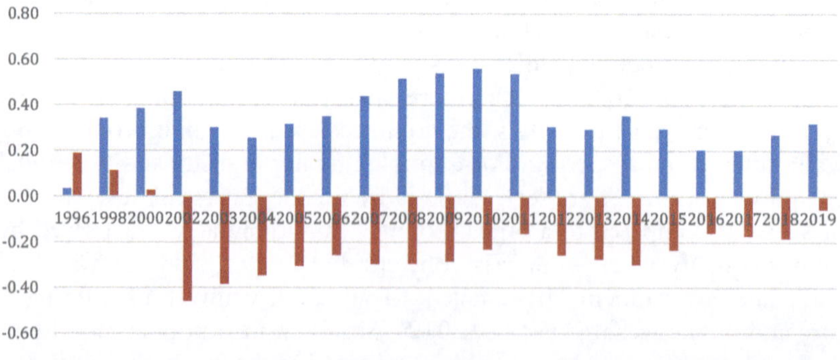

Fig. 9.3 Control of corruption (1996–2019) (*Source* World Bank)

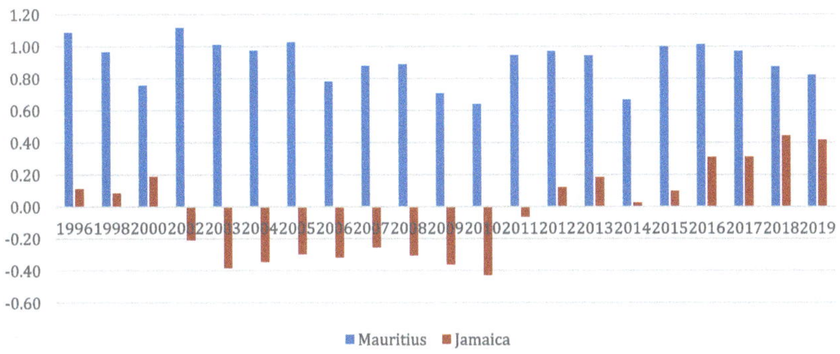

Fig. 9.4 Political stability and absence of violence/terrorism (1996–2019) (*Source* World Bank)

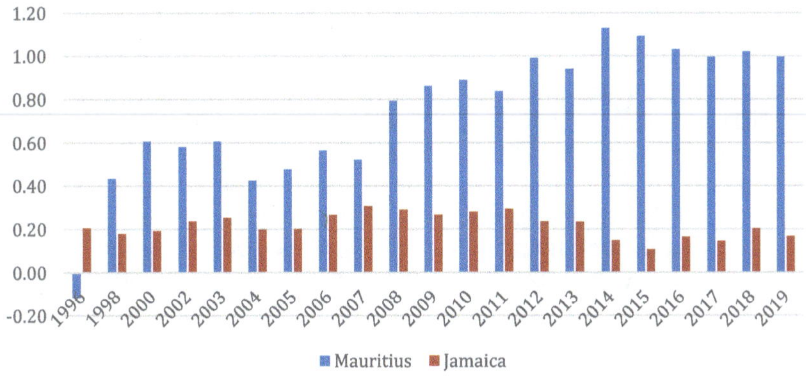

Fig. 9.5 Regulatory quality (1996–2019) (*Source* World Bank)

on all measured institutional fronts including Regulatory Quality, Rule of Law and Government Effective as shown in Figs. 9.5, 9.6 and 9.7, respectively. In the following subsections, a look will be taken at some of the significant features of the Mauritian and Jamaican institutions to see if they can offer some additional explanation for the disparity in growth between the two countries.

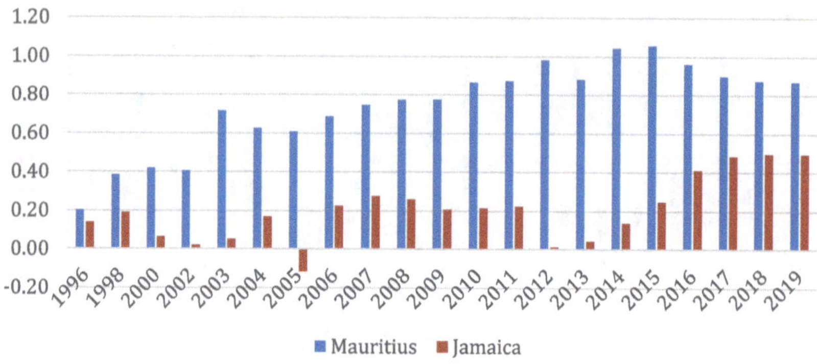

Fig. 9.6 Government effectiveness (1996–2019) (*Source* World Bank)

9.2.2 *What Did Mauritius Do Differently?*

Post-Colonial Mauritius had several key institutions that helped to control corruption on the island (ISS, 2013). These institutions included a strong judiciary, a highly competitive political arena and an independent media. In 2002, Mauritius established the Independence Commission against Corruption (ICAC). This institution is in line with expert advice and international best practices that target three main categories of corruption namely the one-off corruption, the systemic corruption and political corruption (ICAC, 2002). The ICAC has the authority to investigate crimes of corruption and money laundering and prosecute them in court with the consent of the Directorate of Public Prosecution (DPP). Additionally, the ICAC has a preventative and educational role.

Mauritius is ranked at number 52 while Jamaica is ranked at 69th from a total of 180 countries based on the 2020 Corruption Perception Index estimates. Both countries have established institutions to control corruption but Mauritius has been more successful in keeping its corruption level in check relative to Jamaica. Jamaica's failure to prosecute officials when they break the rule of law will be a determinant in the level of progress the country will achieve as sooner or later many other officials will also neglect the rule of law with the perception that they won't be prosecuted. Therefore, the level of corruption that exist in a nation can determine whether the country prospers or suffers from economic retardation, and this is one area that Mauritius has been better than Jamaica which justifies why they have been economically better off as a nation in comparison with Jamaica.

9.2.3 *Jamaica's Inaction*

Jamaica established several institutions over the last forty years to control the level of corruption in the country. These institutions include the Parliamentary Integrity of Members Commission (1975) to police members of parliament assets and liabilities; the Contractor General (1983) to ensure that the government contracts are awarded properly; and the Commission for the Prevention of Corruption (2001) to keep check on the public servants' assets and liabilities. However, concerns have been expressed regarding the adequacy of their impact (NIA, 2013). For example, the Commission for the Prevention of Corruption in its 2012–2013 report expressed that there were concerns regarding the backlog in relations to the rulings of suspected "illicit enrichment". NIA (2013) added that there were cases where the Contractor General requested the prosecution of some public officials but encountered challenge. The political regimes have been rapidly losing their legitimacy, and this has been undermined by their long history of corruption in elections where both the economic and the political elites are implicated in the misuse of funds and other types of corruption (Harriott, 2001). In an atmosphere in which freedom of the press is notable, there is evidence that suggests that this tendency will continue (Powell et al., 2011). Challenges like these are among the reasons why Jamaica has not achieved much economic growth as Mo (2001) posited that a 1% increase in the corruption level reduces the economic growth rate by about 0.72%. Data from the World Bank, in Fig. 9.3, indicates that Jamaica has not been doing enough to challenge corruption. Over a 21-year period (1996 and 2019), Jamaica only has positive score for only 3 years. In essence, 86% of corruption scores over the last 2 decades are negative which provides evidence to suggest that Jamaica has a lot of work to do as a nation to reduce or eradicate corruption. The effect of corruption is far reaching and has many dimensions related to political, economic, social and environmental effects (NIA, 2016). Therefore, the level of corruption that exists in a nation can determine whether the country prospers or suffers from economic retardation. Mo (2001) added that corruption primarily affects economic growth through political instability. Therefore, with Jamaica being politically unstable for the most parts as the data suggest in Fig. 9.8, coupled with high levels of corruption, it is no accident that the economy is growing at such a low rate.

Efforts are still being made to tackle corruption via the National Integrity Action (NIA). The NIA advocates for the passage and enforcement of these and other laws: Single Anti- Corruption Agency (Integrity Commission), Revision of the Political Code of Conduct, Political Party Financing and Registration, Award of Government Contracts, Public Education and Capacity Building among others. However, the NIA does not have the authority to prosecute. There have been cases where senior public officials and a few private sector individuals were brought before the courts, but successful prosecution remains the exception, with only one recent conviction of note (NIA, 2016). Jamaica has also formed the Major Organised Crime and Anti-Corruption Agency (MOCA) in 2014 which brought together the Anti-Corruption Branch (ACB) and the Major Organised Crime and Anti-Corruption Task Force (MOCA TF). This is an elite agency with a focus on tackling corruption in the public sector and bringing high valued targets to justice (MOCA, 2016). Data from the World Bank suggests that the rule of law in Jamaica is not respected, see Fig. 9.7. If the rule of law is violated and corrupted officials do not suffer the consequences, then the institutions established would have been a failure and the country will be in a repetitive cycle characterized by political instability, high levels of corruption and low economic growth which will ultimately be translated into low living standards over time for Jamaicans.

9.2.4 *Post-Independence Nation Building Strategy*

Strong domestic institutions have played a significant role in seeing the Mauritian economy overcoming the macroeconomic imbalances of the 1980s (Subramanian & Roy, 2001). Yao et al. (2005) later added that the high quality of the Mauritian institutions is to be noted when explaining the higher growth rate of Mauritius in relation to both African and other developing nations such as Jamaica. Brutigam (1997) aimed to decipher why political actors in Mauritius abide by the rules that were crafted when similar rules in many other African countries were suspended. He noted

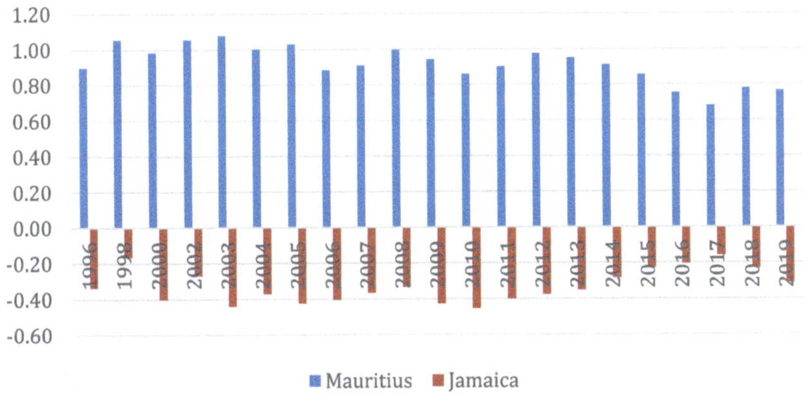

Fig. 9.7 Rule of law (1996–2019) (*Source* World Bank)

that Mauritius was a consolidated democratic[2] state which is marked by institutional features that underpinned the inclusion of major social actors (ethnic groups) in the reform coalition, enabling democratic consolidation to proceed along with economic reform. One such feature of the institution was a rule which determined how the citizens in Mauritius were represented and the structure of the political party system.

The rule which looks at representation and how the party system should be organized was born out of the social tension between different ethnic groups in Mauritius and the fear of being governed by an autocratic government dominated by the Hindu majority. Because of these two factors, 44% of the population (virtually the entire non-Indian population) preferred to remain under British rule and voted against independence (Steer, 2011). In the mid-1960s prior to independence, there were frequent riots including one in which 25 people died (Sriskandarajah, 2005). This led to pessimism regarding the capacity of the country to maintain peace or achieve development. Steer (2011) noted that to ensure legitimacy and to assuage the misgivings of this group (Hindu majority), a nation-building strategy was considered necessary for the post-independence era.

[2] Democratic consolidation is the process by which a newly established democratic regime becomes sufficiently durable that a return to non-democratic rule is no longer likely (Gasiorowski & Power, 1998).

The series of electoral commissions that established the country's electoral system made a careful and deliberate move to organize the democracy in a manner that would not exacerbate the tension between the ethnic groups. All major parties were encouraged to nominate candidates from both the minority and majority communities. Hence, political parties were confident to run in coalitions. Dulal (2013) added that this resulted in institutional arrangements that bolstered the emergence of a democratic developmental state that could ensure economic redistribution and inclusive political representation.[3] There was also a "best losers" system which ensured that all Mauritius' major ethnic groups were represented. Sandbrook (2005) posited that this was where an independent electoral commission appoints up to eight losing candidates to each new National Assembly to represent under-represented ethnic groups. These eight additional seats made it more difficult for a single party to gain the majority of the seats through voting alone since 10% of the seats were reserved (Brutigam, 1997). Since independence in 1968 Mauritius has always been governed by coalition governments of at least two parties, there are instances where more than two political formations shared power (Sithanen, 2003). Thus, by making consensus a necessary feature of the Mauritian government, the rules of the party system ensured that there were no extreme positions on economic policy. This helped to manage economic reform and consolidate the democratic system. From this national consensus, there was a consultative approach towards decision and policymaking rather than a polarization from a single political party. A National Economic and Social Council was established in 2002. It brings together representatives from employers, unions, youth organizations, women's groups, senior citizens, non-governmental organizations, universities and experts in various field. This has been effective in building public support for the policies as well as public trust in the government (Sandbrook, 2005). The nation-building strategy employed by Mauritius enabled a long-term approach to policymaking as well as the facilitation of policy continuity (Steer, 2011). Policy continuity, with the long-term aim to diversify, liberalize and modernize the economy, has played an integral role in building strong business confidence among both domestic and foreign investors (Steer, 2011). Consequently, the single-most important factor explaining the economic success of Mauritius is the

[3] Dulal, H. B. (2013). *Poverty Reduction in a Changing Climate.* Rowman & Littlefield.

achievement of a national consensus on the need to pursue a continuous process of structural adjustment (Government of Mauritius, 2008).

In Jamaica, the structure of the political party system is quite different from the one in Mauritius. Jamaica's party system is characterized by a "first past the post" system. In Caribbean elections, this system discriminates in favour of majority parties and against third parties and minorities (Munroe, 2002). Nations that have adopted a first past the post electoral system usually have a single party, majority government. This can be problematic when consideration is given to long-term policymaking and policy continuation especially when power is exchanged after a political election. For example, in 1989 when the PNP returned to power under the leadership of Micheal Manley, they altered the EPZs' policies that were established by the JLP's Edward Seaga and also attempted to put the EPZ under the control of the Jamaican Port Authority which led to some level of confusion and hindered growth (Hart, 1996).[4] Economic growth depends on good economic policies, but good economic policies are not always in the best interest of political leaders (Bueno de Mesquita et al., 2001). Political leaders in Jamaica have strong incentives to enact policies as election-winning strategies to remain in power, and, as such, often limits the ability for policy continuation if the government should exchange hands. This is a typical behaviour that is stimulated by how the party system is set up which is in contrast with how Mauritius handles their political affairs. It is evident that the Jamaican party system is not set up in a way to accommodate diversity. The establishment of the parliamentary system to accommodate diversity was one of the key institutions that supported the Mauritian miracle (Svirydzenka & Petri, 2014). Therefore, it is not surprising that Mauritius is more politically stable in comparison with Jamaica as suggested in Fig. 9.8. The data shows that Mauritius has been highly stable relative to Jamaica. For most parts, Jamaica can be classified as being politically unstable especially between 2002 and 2011. However, Jamaica has improved since 2012 onwards. More robustly, Mauritius has experienced high levels of stability from 1996 through to 2019.

To facilitate greater political stability and higher institutional quality, it may be worthwhile for Jamaica to pursue a coalition type governmental structure. This would ensure that all political parties whether majority, or

[4] Hart, J. A. (1996). Maquiladorization as a global process. In *Foreign direct investment in a changing global political economy* (pp. 25–38). Palgrave Macmillan.

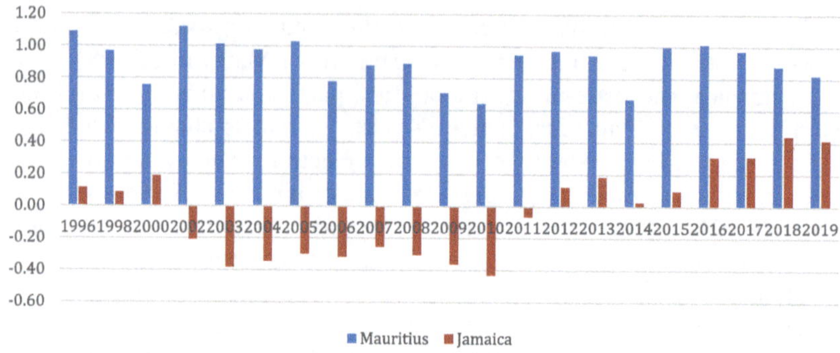

Fig. 9.8 Political stability (1996–2019)

minority, are represented in the government and thus have an input in the economic policies and programmes that are enacted. This would reduce the probability of political opposition and improve political stability and increase the chances of policy continuation in Jamaica which is a positive sign for growth.

9.2.5 Institutions and Trade Policy—Heterodox Opening

Mauritius had a highly restrictive trade regime. This is in contrast with the orthodox schools of economics and the Washington Consensus[5] that believed in a free market and a trade policy of openness. Mauritius chose a strategy of trade liberalization that was unusual and that effectively segmented the export and import competing sectors (Rodrik, 1999). According to Rodrik (1999), through a policy of heterodox opening, Mauritius was able to effectively segment their export sector from the rest of the economy, which prevented the restrictive trade regime from spilling over into this sector. This combination guaranteed that the returns to the export sector remained high enough to prevent domestic resources from being diverted to the inefficient import competing sector (Rodrik, 1999).

The policy of heterodox opening would have probably not been successful without the policies of Mauritius' trading partners which played

[5] The Washington Consensus is a set of economic policy prescriptions considered to constitute the"standard" reform package promoted for crisis-wracked developing countries by institutions such as the International Monetary Fund (IMF) and the World Bank.

an integral role in ensuring the profitability of the export sector. Mauritius has enjoyed preferential access to the markets of the United States and Europe for sugar and textiles that have together accounted for over 90% of the Mauritian exports which is a significant contributor to the economic growth of the country.

Jamaica has alternated between relatively protectionist and liberal policies as a means of managing its trade deficits, safeguarding certain vulnerable productive sectors in the economy, ensuring the availability of foreign exchange and meeting certain political and economic commitments (Hudson, 2006). During the late 1950s, Jamaica's trade policy focused primarily on the use of quantitative restrictions as a means of engendering growth through export promotion and import substitution strategies with a focus protecting designated sectors of the economy in order to promote growth. However, the 1980s marked a significant juncture for Jamaica's trade regime, with the adoption of a more open trade policy for extra-regional trade. Jamaica's structural adjustment of the 1980s which was supported by the IMF and the World Bank saw Jamaica committing not to introduce any new quantitative restrictions on imports. The initial stage of.

Jamaica's Tariff Reform Programme commenced with a 5-year gradual elimination of quantitative restrictions on imports (Hudson & Stennett, 2003).[6] Additionally, many of the original quantitative restrictions were removed between 1982 and 1983, while some were converted to tariffs with the aim of progressively lowering the rates. The desired effect was to expose production for the domestic market to limited external competition, thereby encouraging domestic producers to be more competitive, while maintaining some level of protection for the local economy. However, in the context of Jamaica's commitments under the WTO and future negotiations with the EU and the prospective FTAA member states, it is likely that trade liberalization, specifically tariff reductions, in Jamaica will accelerate (Hudson, 2006).

This is typical of Orthodox economic policies and a contrast to Mauritius developmental heterodox opening strategy.

[6] https://cert-net.com/files/publications/conference/824.pdf.

9.2.6 *Export Processing Zones*

The benefits derived from the Export Processing Zones (EPZs) are to be credited to strong domestic institutional quality. Other EPZs established by Jamaica and other African nations provide the same incentives for investors as the zones in Mauritius. The difference is that the zones in those countries are hampered by inefficient bureaucracy, rent seeking abuses, corruption and political opposition (Vastveit, 2013). Madani (1999) explains that the EPZ Law was born out of the failure of Import Substitution Industrialisation (ISI).[7] The reasons for the failure of the ISI strategy to diversify the economy by developing manufacturing industries and thus a new structure of production were due to a lack of capital, shortage of skilled workers and lack of enterprise and risk-taking regarding new industrial activities (most probably due to the deeply ingrained sugar mentality in the country) (YeungLamKo, 1998). Therefore, the Government of Mauritius had to act very quickly after coming to the realization that the ISI strategy was not as effective as they thought it would have been to solve the main issue of unemployment in the country. The government then switched its development strategy to an export-orientated industrialization (EOI) one where a legislation which embodied the Export Processing Zone (EPZ) Act No. 51 of 1970, an Export Processing Zone, or a Free Zone as it is also sometimes called. Almost all the EPZ factories established after 1971 were in the textile and apparel industry where females were employed at low wages. The incentives for the EPZs included exemption from excise and duties on productive machinery and parts; exemption from payment of income tax on dividends and profits for the first ten years (*Arthur Lewis: Economic Development with unlimited Supply of labour* [Lewis, 1954]); availability of factory buildings and fully serviced land at reasonable or subsidized rates; preferential interest rates and favourable labour laws (dismissal and overtime); and all production must be exported. Foreign investors also enjoyed the following conditions: guarantee against nationalization; free repatriation of capital (except capital appreciation), profits and dividends; and the availability of permanent residence and work permits for foreign technicians and managers.

[7] A trade and economic policy which advocates replacing foreign imports with domestic production. ISI is based on the premise that a country should attempt to reduce its foreign dependency through the local production of industrialized products policies.

For the period 1985–1995, the economy grew at an average of more than 5.0% annually in real terms, with a significant contribution coming from the EPZ sector which also increased employment (Ramasamy, 2014). Employment in the EPZ firms helped to reduce the unemployment rate from approximately 20% to under 3% in the late 1980s. The EPZs employed more than 90,000 people in approximately 600 factories in 1990 and exports reached $738 million in 1990 which was about 60% of the total exports of the country and approximately 35% of the gross domestic product (Chan, 2016). After 1988, both labour and capital accumulation stopped, but output kept growing. This was a period of pure productivity-driven growth (Subramanian & Roy, 2001). Mauritius had enjoyed total factor productivity growth which has come primarily from the EPZs exports. Chernoff and Warner (2002) noted that there was a gradual process of learning that took several years to bear fruit. The TFP growth started after new vintage capital was put in place in the EPZs in the mid-1980s and has continued uninterrupted through the end of the 1990s.

Figure 9.9 illustrates a convergence of TFP growth rates between Jamaica and Mauritius. This converge shows a gradual decline in Jamaica's performance relative to a steady increase in Mauritius's. The convergence started in the early 1970s when the EPZ law was passed in 1971 in Mauritius where many unproductive females entered the labour force via firms

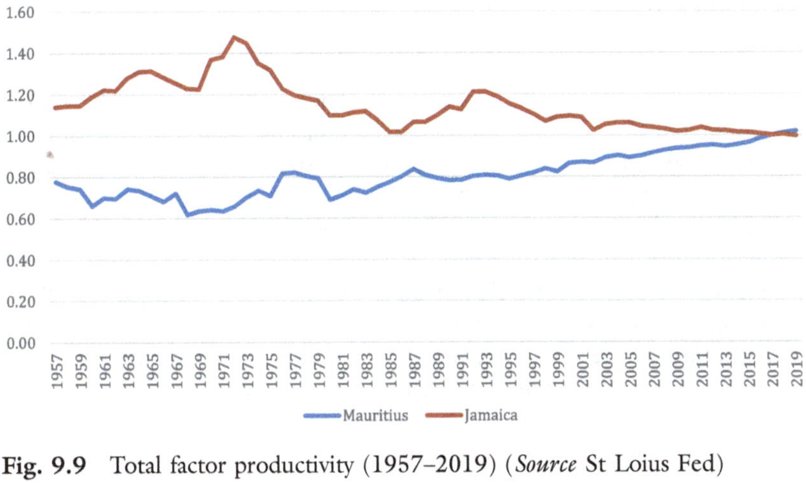

Fig. 9.9 Total factor productivity (1957–2019) (*Source* St Loius Fed)

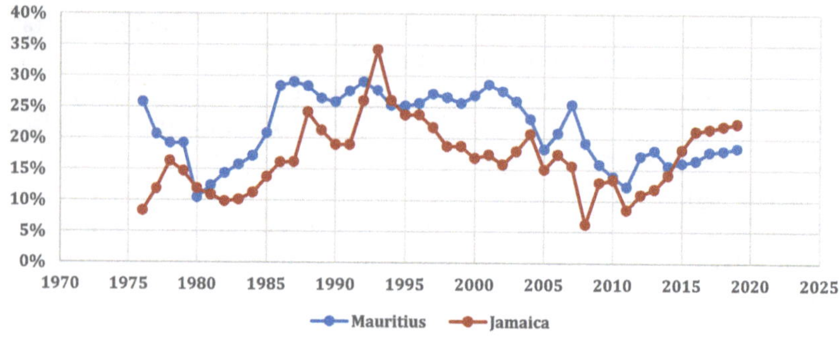

Fig. 9.10 Gross savings (1976–2019) (*Source* World Bank)

in the EPZs and were now adding to Mauritius productivity in exports. On the other hand, a global oil price shock resulted in a year-long crisis in 1972 which underpinned the Jamaica economy and placed it on a path to trade and fiscal imbalances and lower TFP growth. The disparity in TFP between the two countries was reduced from 0.75 percentage points in 1973 to 0.23 percentage point in 1987 before facing another external shock in hurricane Gilbert in 1988. Hurricane Gilbert which is arguably one of most destructive natural disaster to have impacted Jamaica. In summary, the EPZ in Mauritius has been one of the key factors that have contributed to the economic growth of Mauritius and in extension the *Mauritian Miracle*.

Jamaica's structural adjustment in the 1970s was marked by the removal of exchange rate controls, the removal of tariffs and quotas and the removal of subsidies from farmers. These were all geared towards preparing Jamaica for the neo-liberal economic environment and to propel the country into the global economic space. Figure 9.10 illustrates that Jamaica's savings rate has been low since the 1970s.

The Neo-classical, Harrod-Domar Model[8] suggests that if developing countries want to achieve economic growth, governments need to encourage savings. In macroeconomic theory, savings is equal to investments. Therefore, if a country is not saving enough, they will not have enough money to invest and will have to seek investments overseas via borrowing. In Jamaica at that time, Foreign Direct Investments

[8] See Harrod (1939) and Domar (1946).

(FDI) came in the form of the EPZs because Jamaica's savings' rate was relatively low relative to Mauritius in the 1970s, see Fig. 9.10. The Jamaican EPZs were created in 1976 under the Edward Seaga Government (Chan, 2016). This was in line with Lewis (1954) approach of *Economic development with unlimited supplies of labour*. There were three zones: Kingston EPZ (1976) which began with 140 employees (WBIEDID, 1992), Montego Bay Free (1985) and Garmex free zone in 1988. Kingston EPZ grew rapidly until 1981 when growth slackened off due mainly to the economic recession in the United States. After 1983, growth resumed until another downturn in 1988 which was induced by the combination of a destructive hurricane and the labour militancy zone (Chan, 2016). In 1989, when the People's National Party (PNP) came out victorious by winning 45 of the 60 seats in the general elections, Michael Manley returned to power (EOJ, 2016). Manley altered the EPZ policies and made attempts to put the EPZs under the control of Jamaica port authority which created some confusion within the EPZs community about the future of the state policies (Hart, 1996).[9]

The EPZs were not diversified and only concentrated on textiles. The employees of the EPZs were involved in very low skilled jobs where they earned very low wages similarly to the Mauritian case. Chan (2016) noted that the main attractions for foreign investors were low wages and the proximity of Jamaica to US markets. The conditions in which the workers worked were unfavourable and trade unionists tried to negotiate with the employers but were blocked initially. However, by the early 1990s when globalization expanded and the trade unions managed to bargain with the employers, the firms in the EPZs began to leave Jamaica one after the other. This was possible since globalizaiton facilitated the free movement of capital and labour, so businesses could afford to leave and venture to other countries where in their view had a more stable labour environment and possibly even cheaper labour.

The North American Free Trade Agreement (NAFTA) between Canada, Mexico and the United States was also a contributor to the failure of the EPZs in Jamaica. This agreement encapsulated an open market objective that was consistent with GATT[10] regulations. It stated

[9] Hart, J. A. (1996). Maquiladorization as a global process. In *Foreign direct investment in a changing global political economy* (pp. 25–38). Palgrave Macmillan.

[10] The General Agreement on Tariffs and Trade covers international trade in goods. The workings of the GATT agreement is the responsibility of the Council for Trade

that all tariffs must be eliminated within the free trade area over a transition period (Lassanyi, 1992). Along with tariff elimination, NAFTA included a reduction in barriers to trading in textiles and apparel among other things. The reduction of the barriers regarding textiles and apparel made it easier cheaper for US firms to set up business in Mexico and then transport goods to the United States. NAFTA made trading between Mexico, United States and Canada more attractive relative to other markets like Latin America and the Caribbean. There was no evidence to suggest that governments within these regions made efforts to counter balance NAFTA by minimizing their own trade barriers at the time. This weakened the market in Jamaica and ultimately contributed to the failure of the EPZ there. Vastveit (2013) pointed out that there were other EPZs that provided the same incentives as the zones in Mauritius but were not as fruitful as Mauritius.

The EPZs in Jamaica did not contribute as much to the growth of exports, employment and GDP as the EPZs in Mauritius. Chan (2016) noted that this result was not surprising since there were ambivalences in the government policies and uncertainty about the continuity of those policies in Jamaica. On the contrary, Mauritius had a coalition government that minimized any political oppositions. As Vastveit (2013) pointed out, a major contributing factor to the success of the EPZs in Mauritius was because the institutions were strong, and the country did not suffer from the political opposition as Jamaica did.

Jamaica has always lacked the physical infrastructure and human capital base to move beyond the simple assembly operations to more sophisticated manufacturing processes. In Mauritius, availability of factory buildings and fully serviced land at reasonable or subsidized rates was an incentive to investors (YeungLamKo, 1998). Thus, Jamaica's experience was less successful than that of Mauritius.

Today, Jamaica can still be considered as having EPZs. This is in the form of the ever-expanding Global Services Sector (GSS). According to the Jamaica Information System (JIS), Jamaica has attracted over 60 companies to the industry up to July in 2021.[11] The GSS has made a significant contribution to the labour market of the Jamaican economy by

in Goods (Goods Council) which is made up of representatives from all WTO member countries (WTO, 2016).

[11] JIS (July, 2021). Global Services Sector Project Progressing Well. https://jis.gov.jm/features/global-services-sector-project-progressing-well/.

creating over 40,000 jobs as well as adding approximately US$ 600,000 million in revenues on an annual basis. In the early stages of the development of the GSS in the country, jobs were predominately created in Business Process Outsourcing (BPO) firms which did not require a very high skill level. In this form, the GSS may not have been contributing significantly to the development of the island which already has employment growing faster than output—a recipe for declining productivity growth. However, there have been strategies to upgrade or upskill within the sector. This has led to the creation of Knowledge Process Outsourcing (KPOs), LPOs and other types of businesses. This will also be integral for Jamaica to integrate itself in the multi-billion-dollar IT and KPO global markets. The IT Outsourcing market is projected to reach US $425.2 billion by 2026, while the KPO market is estimated to have a value of US $ 124.3 billion by 2025.[12] If Jamaica capitalizes on these opportunities, then the country can benefit in a major way. Notwithstanding the benefits to be derived from the GSS sector, the associated risks are similar to the EPZs of the 1980s. Has Jamaica learn any lessons from that experience? Is there anything to hinder these firms from moving their businesses to other cost-effective locations if the opportunity presents itself? How will the over 40,000 Jamaicans working in these sectors support themselves and their families if the firms should leave? Are Jamaicans in these firms adequately trained to operate within the industry by themselves or know how to position themselves to receive contracts to operate a GSS firm? These are the many questions that needs addressing to determine the future viability of this sector and not repeating the EPZ experience.

9.2.7 Culture and Ethnicity

Culture is characterized as those customary beliefs and values that ethnic, religious and social groups transmit unchanged from generation to generation (Guiso et al., 2006). A country's culture includes the values, preferences and trust which shapes the attitude of its people towards work which ultimately determine its economic performance (Abramovitz, 1995). Therefore, differences in culture can have different implications for economic growth.

[12] Skill Upgrade Strategy being finalized for Global Services Sector https://jamaica.loopnews.com/content/skills-upgrade-strategy-being-finalised-global-services-sector.

Ramasamy (2014) noted that there were three different ethnic groups coexisting in Mauritius. The group consisted of the Indo-Mauritians,[13] General population[14] and, finally, the Sino-Mauritians.[15] Approximately 68% of the Mauritian population are Indo-Mauritian, 27% Creole, 3% Sino-Mauritians and 2% Franco- Mauritian.[16] Even though the Sino-Mauritians (Chinese) were a very small proportion of the Mauritian population, they played a significant role in the success of the Mauritian miracle. The Chinese have a business-oriented culture and are always seeking to set up shops or establish some form of business wherever they settle. Sino-Mauritian business networks helped ensure that in the first decade of operation, some 60% of the capital investment in the EPZ came from Hong Kong and rest from Taiwan, Malaysia and Singapore (Brautigam, 2005). The first industrialist who came exclusively from China was the Esquel Group (Ramasamy, 2014). The Esquel Group is one of the world's largest producers of cotton shirts which has their main in China, Malaysia, Vietnam, Mauritius and Sri Lanka (Cao et al., 2008).

The Chinese arrived in Mauritius, while the country was still under the leadership of the French where they set up shops and occasionally planted tea. When the British took over from the French, there was an increase in immigration, and by 1850, the Chinese dominated the small shops in Port Louis and the rural areas. Chinese Chamber of Commerce was organized in 1908, and the Chinese began to establish some of the country's first factories before and after the turn of the century. Romer (1990, 1993) contributed an approach to growth theory based on ideas (innovations in either products or production methods) as the key ingredient for development, rather than capital, labour or other factors of production. Romer (1992) also added that importing ideas from abroad, through inward FDI, is an effective alternative to growing them at home. This was evident in Mauritius as Chinese businessmen brought the idea of textile and apparel manufacturing to the EPZ which jump-starting the country's industrialization (Frankel, 2014).

[13] A combination of Hindus and Muslims.

[14] Mainly of people of African descent (creoles) and the rest Franco- Mauritians (French origin) and coloured.

[15] Trace their ethnic ancestry from China.

[16] Data collected from https://www.mauritiusinsideout.com/population-of-mauritius.html.

The Chinese population maintained contacts with their ancestral villages in mainland China for family visits and to bring spouses back to Mauritius. As other villagers settled in other parts of Asia, family networks expanded to connect Mauritian Chinese to their kin throughout Asia. With the advent of communist rule in China, many of these relatives fled to Hong Kong and Taiwan, which made family connections particularly strong in those two areas. Although Asian investment has declined over the past decade, as of 2001 the largest foreign-owned firms and joint ventures in the EPZs were still those established by investors from Hong Kong, for example, Novel Textiles, Sentosa Enterprises, the Esquel Group, Sinotex, Goodlands and Star Knitwear, among others. The establishment of the EPZ in 1970 and its subsequent success is directly tied to these Asian networks (Brautigam, 2005). Ramasamy (2014) noted that after the Sino-Mauritian diaspora who were the pioneers of the EPZs, some Franco-Mauritians sugar barons[17] and the Indo-Mauritians also became interested in the textile export industry. The Currimjee Group with Bonair Industries were the first Indo-Mauritians to enter the textile export industry. It is interesting to note that they were already operating under Development Certificates where they produced soap. There were others who owned DC firms under the ISI strategy but also invested in the EPZs. They had the benefit of being already familiar with industrialization and factory work which made it relatively easier for them to diversify in the EPZ in comparison with those who did not have prior knowledge or experience in industry.

The Jamaica situation is a bit similar but vastly different in outcome. There are also different ethnic groups coexisting in Jamaica, the majority of whom are of African descent. It is estimated that 76.3% are of African descent, 15.1% Afro-European, 3.4% East Indian and Afro-East-Indian, 3.2% Caucasian, 1.2% Chinese and 0.8% from other groups.[18] This is a direct result of slavery where blacks were transported from Africa to the Caribbean to work as slaves on sugar plantations. Chinese and Indians later came to Jamaica as indentured labourers. After completing their period of indentured labour, they moved into business and other professions, and although small in number, they still have significant influence

[17] The Dalais family created Floreal Knitwear, Tim Taylor with Cobra and Francois de Grivel with Festival.

[18] Data extracted from https://worldpopulationreview.com/countries/jamaica-population.

in Jamaica in these areas. However, the impact of the Chinese on the economic growth of Jamaica was not as significant as the Chinese in Mauritius due to various factors. Prime among which is that the Chinese and other Asians were not as sentimental to Jamaica as they were Mauritius. They were not involved in the political continuity in Jamaica as they were in Mauritius, and they mostly repatriated most of their profits from Jamaica. A large share of the work permits that are granted to foreigners in Jamaica goes to the Asians. Along with the Asians who have lived in Jamaica for generations, there are other who come on an annual basis seeking permits to work in Jamaica. In 2019, more than half (52%) of the work permit recipients were Asian. In 2020, there was a reduction in the number of work permits granted which is likely to have been influenced by COVID-19. However, the Asians still accounted for the greatest share of permits (56.5%) granted by the Jamaican. While they are largely involved in wholesale and retail trade activities, they are less involved with the political landscape and as a result do not have the same impact as they have had in Mauritius.

9.3 Conclusion and Policy Prescriptions

The *Mauritian Miracle* was as a result of an amalgamation of good institutions, a coalition government, the establishment of Export Processing Zones, Sino-Mauritian influences and also heterodox approach to trade policy. However, the institutions might have been the most significant contributor to the overall development of the country. It was due to good institutions that the EPZ worked because as noted by Vastveit (2013), there were other EPZs that provided the same incentives as the zones in Mauritius, but those zones were hampered by rent seeking abuses, corruption and political oppositions. Due to the establishment of a coalition government, political opposition was kept low which also increased the prospect for long-term policy planning and policy continuation. Mauritius also established the ICAC which was an anti-corruption agency which tackled one-off, systematic and political corruption. This agency has the power to prosecute corrupt public officials which serve as a deterrent to corrupt practices. All these factors have contributed to the economic growth in Mauritius. Jamaica, on the other hand, has not managed to match the economic growth of Mauritius. Jamaica has suffered from institutions' weakness, as noted by Duncan-Waite and Woolcock (2008), corruption and political oppositions which are in stark contrast with Mauritius. Jamaica had set up EPZs from the 1970s as Mauritius did, but it was not as successful. However, some aspects remain in Jamaica today and there are useful lessons from the initial EPZs that can be learnt. Based on the findings of the major determinants of the success of the Mauritian Miracle, the following policies can be applied to Jamaica which might be useful in transforming Jamaica from a third world to a first world state:

1. Coalition Government

    - Jamaica has suffered from the problem of policy continuity leading back to the time of Edward Seaga and Michael Manley in the 1970s and 80s. Leaders often focus on policies that would act as election-winning strategies which often undermined good economic policy and economic growth in Jamaica. After a political election, policies that were enacted while the opposing political party was in power were often removed or altered if a new government was elected, so the goal of those policies was not achieved even if it would have been fruitful if it had lasted its full tenure. This tends to waste valuable resources. However, with a coalition government, the major political parties can be represented in the government as well as minority parties where there can be a consensus on economic policymaking. This would reduce the extent of political opposition that Jamaica has been known for and also increase the probability of policy continuation. The IMF has recently made a set of specific guidelines for Jamaica while the PNP was in power and because of those guidelines the JLP saw those policies through which has brightened the macroeconomic outlook for Jamaica.

2. Export Processing Zones

    - Jamaica's initial experience with EPZs was not fruitful for the Jamaican economy. In the 1970s, Jamaica required FDI inflows which came in the form of the EPZs, but the negotiations were unfavourable to Jamaica and ultimate when the firms left, there were no benefits to be derived. Today, some aspect of EPZs are still evident in Jamaica in the form of BPOs. These jobs that are being outsourced are of a low skilled level and require a shift to more sophisticated jobs which require innovation and critical thinking to propel the Jamaica economy forward. This policy is in line with new growth development theorist like Lucas (1988) who suggest that by doing jobs that force one to think, or *learn by doing*, the stock of knowledge and human capital will increase. Now it is perfect for a move like this as the Panama Canal has been enlarged and will require lawyers and accountants to operate in legal and financial services among others.

3. Caymanas Special Economic Zone

   - The Special Economic Zone must be diversified in order to ensure that there are multiple income streams for the economy so that a single external shock or a scarce factor does not push the economy in an extensive downturn.
   - Foreign investors must be tied to long-term contracts which a specific law to a partnership with local firms to ensure that Jamaica benefits from any contracts that are being negotiated. This is to ensure that the experience of the EPZs of the 1970s is not repeated in the context of the Special Economic Zones.

4. Anti-Corruption Agency

   - Jamaica has established a series of anti-corruption agencies, but these agencies lack the power to prosecute corrupted public officials. There have been cases where the contractor general requested prosecution of several public officials which has proven futile. If this continue, then the problem of corruption will be exacerbated and affect Jamaica's growth negatively. Therefore, there is the need for an anti-corruption agency with the power to prosecute as the ICAC in Mauritius.

References

Abramovitz, M. (1995). The elements of social capability. *Social capability and long-term economic growth* (pp. 19–47). Macmillan.

Acemoglu, D., Johnson, S., & Robinson, J. A. (2005). Institutions as a fundamental cause of long run growth. *Handbook of Economic Growth, 1*, 385–472.

Adams, R. H. (2003). *Economic growth, inequality and poverty: Findings from a new data set* (Vol. 2972). World Bank Publications.

Bhalotra, S. R., et al. (2006). *Childhood mortality and economic growth.* United Nations University.

Birdsall, N., Rodrik, D., & Subramanian, A. (2005). How to help poor countries. *Foreign Affairs, 84,* 136–152.

Blavy, R. (2006). *Public debt and productivity: The difficult quest for growth in Jamaica* (Nos. 2006–2235). International Monetary Fund. Retrieved from https://books.google.com.jm/books?id=jSRR4TpIYnMC

Brautigam, D. A. (2005). Strategic engagement: Markets, transnational networks, and globalization in mauritius. *Yale Journal of International Affairs 1,* 63.

Brutigam, D. (1997). Institutions, economic reform, and democratic consolidation in mauritius. *Comparative Politics, 30*(1), 45–62. Retrieved from http://www.jstor.org/stable/

Bueno de Mesquita, B., Morrow, J. D., Siverson, R. M., & Smith, A. (2001). Political competition and economic growth. *Journal of Democracy, 12*(1), 58–72.

Cao, N., Zhang, Z., Man To, K., & Po Ng, K. (2008). How are supply chains coordinated? An empirical observation in textile-apparel businesses. *Journal of Fashion Marketing and Management: An International Journal, 12*(3), 384–397.

Chan, S. (2016). *Foreign direct investment in a changing global political economy.* Palgrave Macmillan UK. Retrieved from https://books.google.com.jm/books?id=oNu-DAAAQBAJ

Chernoff, B., & Warner, A. (2002). *Sources of fast growth in mauritius: 1960–2000.* Harvard University.

Dhingra, S., Ottaviano, G. I., Sampson, T., & Reenen, J. V. (2016). *The consequences of brexit for uk trade and living standards.*

Domar, E. D. (1946). Capital Expansion, Rate of Growth, and Employment. *Econometrica, 14*(2), 137–147. https://doi.org/10.2307/1905364

Duncan-Waite, I., & Woolcock, M. (2008). *Arrested development: The political origins and socio-economic foundations of common violence in Jamaica* (Working Paper [46]). Brooks World Poverty Institute.

EOJ. (2016, November). *General elections 1989.* Retrieved from http://www.eoj.com.jm/content-70-221.htm

Factbook, C. (2015). *The world factbook,* 10. CIA–Central Intelligence Agency. Dispon´ıvel em:https://www.cia.gov/cia/publications/factbook/Acesso/em.

Ferrini, L. (2012). *The importance of institutions to economic development.* E-International Relations Students.

Frankel, J. (2014). Mauritius: African success story. In *African successes: Sustainable growth* (Vol. 4). University of Chicago Press.

Gasiorowski, M. J., & Power, T. J. (1998). The structural determinants of democratic consolidation evidence from the third world. *Comparative Political Studies, 31*(6), 740–771.

Government of Mauritius. (2008). *Mauritius 40 years after: New goals, new challenges—Where it all started*. Republic of Mauritius. Retrieved from https://books.google.com.jm/books?id=czsAQAAIAAJ

GOVMU. (2005, September). *A roadmap for the mauritius sugarcane industry for the 21st century*. Retrieved from http://agriculture.govmu.org/English/Documents/Archives/Policy%20Documents%20and%20Reports/roadmap.pdf

Guiso, L., Sapienza, P., & Zingales, L. (2006). Does culture affect economic outcomes? *The Journal of Economic Perspectives, 20*(2), 23–48.

Harrod, R. F. (1939). An essay in dynamic theory. *The Economic Journal, 49*(193), 14–33. https://doi.org/10.2307/2225181

Harriott, A. (2001). The crisis of public safety in Jamaica and the prospects for change. *Souls, 3*, 56–65.

Helpman, E. (2005). *The mystery of economic growth*. Academic Foundation. Retrieved from https://books.google.com.jm/books?id=lcPm0nRIITwC

HM Treasury, D. (2005). *From commitment to action: Human development and growth*.

Hudson, S. (2006). Trade liberalisation and the Jamaican economy: Prospects and effects of tariff adjustments. In A. Birchwood & D. Seerattan (Eds.), *Finance and real development in the Caribbean* (pp. 498–528). Caribbean Centre for Monetary Studies. The University of the West Indies ; Republic of Trinidad and Tobago.

Hudson, S., & Stennett, R. (2003). Current account sustainability in Jamaica. *Bank of Jamaica, Working Paper, 2*(11), 1–43.

ICAC. (2002, February). *Independent commission against corruption (icac)*. Retrieved from http://www.iaaca.org/AntiCorruptionAuthorities/ByCountriesandRegions/M/Mauritius/201202/t20120210802416.shtml

IMF. (2003, April). *World economic outlook: Growth and institutions*. Retrieved from https://www.imf.org/external/pubs/ft/weo/2003/01/pdf/front.pdf

ISS. (2013). *From a rocky start to regional leadership: Mauritius's—Anti-corruption agency, 2006–2012*. Princeton University.

Lassanyi, M. (1992). *North American free trade agreement (nafta): Mexico*. Diane Publishing Company. Retrieved from https://books.google.com.jm/books?id=bef6VE2RXYEC

Lewis, W. A. (1954). Economic development with unlimited supplies of labour. *The Manchester School, 22*(2), 139–191.

Lin, B. Q. (2003). Economic growth, income inequality, and poverty reduction in people's Republic of China. *Asian Development Review, 20*(2), 105–124.

Lucas, R. E. (1988). On the mechanics of economic development. *Journal of Monetary Economics, 22*(1), 3–42.

Madani, D. (1999). *A review of the role and impact of export processing zones* (Vol. 2238). World Bank Publications.

Mo, P. H. (2001). Corruption and economic growth. *Journal of Comparative Economics, 29*(1), 66–79.
MOCA. (2016, November). *Major organised crime and anti-corruption agency: About us*. Retrieved from http://www.moca.gov.jm/
Mundi, I. (2016, November). *Mauritius demographics profile 2016*. Retrieved from http://www.indexmundi.com/mauritius/demographicsprofile.html
Munroe, T. (2002). *An introduction to politics: Lectures for first-year students*. Canoe Press. Retrieved from https://books.google.com.jm/books?id=JdiLyC1asTcC
NIA. (2013, April). *National integrity action: Why Jamaica urgently needs an effective integrity commission*. Retrieved from https://niajamaica.org/issues/campaign-financing/22-issues/901-why-jamaica-urgently-needs-an-effective-integrity-commission
NIA. (2016, November). *Our priority issues*. Retrieved from https://niajamaica.org/issues/campaign-financing
Nordea. (2021, March). *Mauritius: Economic outline*. Retrieved from https://www.nordeatrade.com/dk/explore-new-market/mauritius/economy
North, D. (1990). *Institutions, institutional change and economic performance*. Cambridge University Press. Retrieved from https://books.google.com.jm/books?id=oFnWbTqgNPYC
Powell, L. A., Lewis, B., & Seligson, M. (2011). *Political culture of democracy in Jamaica, 2010: Democratic consolidation in the Americas in hard times*. USAID.
Ramasamy, J. K. (2014). *School of economic and business sciences faculty of commerce, law and management the university of the witwatersrand* (Unpublished doctoral dissertation). The University of the Witwatersrand.
Ravallion, M. (2005). *Inequality is bad for the poor* (World Bank Policy Research Working Paper [3677]).
Rodrik, D. (1999). *The new global economy and developing countries: Making openness work* (Vol. 24). Overseas Development Council.
Rodrik, D. (2008). *One economics, many recipes: Globalization, institutions, and economic growth*. Princeton University Press.
Romer, P. (1990). *Endogenous technological change* (Tech. Rep.). National Bureau of Economic Research.
Romer, P. (1992). Two strategies for economic development: Using ideas and producing ideas. *The World Bank Economic Review, 6*(suppl 1), 63–91.
Romer, P. (1993). Idea gaps and object gaps in economic development. *Journal of Monetary Economics, 32*(3), 543–573.
Sandbrook, R. (2005). Origins of the democratic developmental state: interrogating mauritius. *Canadian Journal of African Studies/La Revue canadienne des 'etudes africaines, 39(*3), 549–581.

Sithanen, R. (2003). *Coalition politics under the tropics: Office seekers, power makers, nation building—A case study of mauritius* (Vol. 19). EISA Roundtable on Political Party Coalitions.

Sriskandarajah, D. (2005). Development, inequality and ethnic accommodation: Clues from Malaysia, Mauritius and Trinidad and Tobago. *Oxford Development Studies, 33*(1), 63–79.

Steer, L. (2011). *Progress in economic conditions*. Overseas Development Institute Publications, 111 Westminster Bridge Road, London SE1 7JD, UK.

Subramanian, M. A., & Roy, M. D. (2001). *Who can explain the mauritian miracle: Meade, Romer, Sachs, or Rodrik?* (No. 1–116). International Monetary Fund.

Sugar Industry Authority of Jamaica. (2000, April). *An overview of the sugar industry of Jamaica* Retrieved from http://www.jamaicasugar.org/SIASection/Overview.pdf.

Svirydzenka, K., & Petri, M. M. (2014). *Mauritius the drivers of growth—Can the past be extended?* (No. 14–134). International Monetary Fund.

UNDP. (2020, May). *Social and economic impact of the covid-19 and policy options in Jamaica. United nations development programme*. One United Nations Plaza.

Vastveit, L. K. (2013). *Export processing zones in Sub-Saharan Africa–Kenya and Lesotho.*

WEF. (2019). *The global competitiveness report 2019*. Retrieved from http://www3.weforum.org/docs/WEF_TheGlobalCompetitivenessReport2019.pdf

WBIEDID, W. D., Division. (1992). *Export processing zones* (No. 75). World Bank. Retrieved from https://books.google.com.jm/books?id=xJBoyutM4KMC

World Bank, T. (2016a, October). *Beyond economic growth student book*. Retrieved from http://www.worldbank.org/depweb/english/beyond/global/glossary.html

World Bank, T. (2016b, October). *Overview of Jamaica*. Retrieved from http://www.worldbank.org/en/country/jamaica/overview

WTO. (2016, November). *Gatt and the goods council*. Retrieved from https://www.wto.org/english/tratope/gatte/gatte.htm

Wynter, B. (2011, September). *Address to the Jamaica chamber of commerce luncheon meeting: Productivity in Jamaica*. Retrieved from http://boj.org.jm/uploads/news/speech-jccluncheon,2011-9-13.pdf

Yao, J., El-Masry, G., Khandelwal, P., & Sacerdoti, E. (2005). *Mauritius: Challenges of sustained growth*. International Monetary Fund. Retrieved from https://books.google.com.jm/books?id=in-GjMBuNPIC

YeungLamKo, L. (1998). *The economic development of mauritius since independence*. University of New South Wales.

CHAPTER 10

Forging a Way Forward

Economic development requires the efficient coordination of a country's resources with minimal inefficiency and wastage. At the heart of this locomotion is its people whose labour can be repositioned from low productive to highly productive inputs. Through economic transformation, a country has the potential to increase the value of the goods and services it sells to the rest of the world and increase its income which reduces how much it has to borrow. Much is available to invest in infrastructure and social services, for example education, health care and crime prevention. In 2007, Jamaica launched its plan to achieve developmental status of US $12,535 in Gross National Income (GNI) per person by 2030. At that time, Jamaica's GNI was roughly $7,830 per person, and therefore, the country only needed little under $5000 to achieve the goal. Since then, Jamaica has regressed. At the end of 2022, GNI per person fell to approximately US $4,800, a 40% decline instead of the needed 40% increase. Now the country is further away from its target instead of advancing towards it.

Equation 10.1 calculates the time in years that it might take Jamaica to earn the requisite GNI per capita based on historical growth rates, holding all other factors constant. Here T is time in years, LnA_E is the

A. Haughton and W. Ivey, *Overcoming Productivity Challenges in Small Countries*, https://doi.org/10.1007/978-3-031-23301-2_10

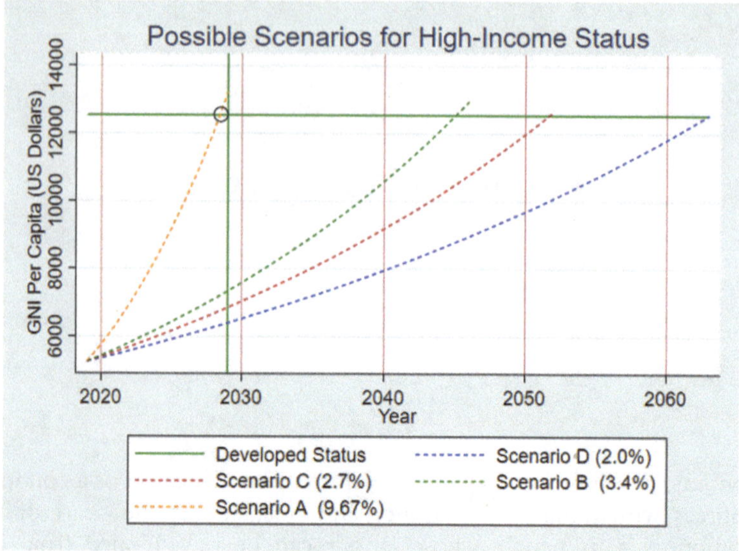

Fig. 10.1 High income status scenarios (*Source* World Bank)

predicted level of output, LnA_1 the current level of output and r represents the growth rate (discount factor). The different resulting scenarios are illustrated in Fig. 10.1.

$$T = (LnA_E - LnA_1)/r \tag{10.1}$$

Jamaica's GNI per capita will have to grow by roughly 9.67% per annum to achieve developmental status by 2030 as illustrated in Scenario A. This is highly unlikely given that historically, Jamaica's GNI grew by only 3.4% pre Global Financial Crisis (2000–2009), 2% post GFC (2010–2019) and 2.7% on average for both decades combined (2000–2019). Scenario B shows that if GNI grows by the pre GFC average of 3.4%, the goals can be achieved by 2045. Scenario C shows that if GNI grows by 2.7%, Jamaica could become developed by 2052. Finally, in Scenario D it is projected that if Jamaica's GNI were to grow by 2%, its post GFC average, the country could achieve its developmental goal by 2060. Jamaica's Medium Term Socioeconomic Plan reveals that only 27% or three quarters of the developmental targets were met for the 2009–2018 period.[1] Thirty per cent (30%) improved over the baseline but did not meet the target, 36% was either worse or equal to the 2007 baseline and 7% could not be measured. This assessment was conducted before the onset on the COVID-19 pandemic in 2020. As explained in Chapter 1, Jamaica faces an insurmountable challenge especially given the impact of the COVID-19 pandemic, a recession that has been exacerbated by global geopolitical occurrences, increased oil prices and global inflation. It is likely that some of these targets may fall further behind. However, if GNI growth is to increase and Jamaica is to become a developed country, productivity growth is of utmost importance.

Consistent economic growth depends on consistent productivity growth (Timmer et. al, 2011).[2] Therefore, targeting productivity growth is as important element of targeting economic growth. Along with overcoming the challenges highlighted in Chapters 1 to 9 including low levels of human capital, intensified crime and violence, institutional weaknesses, corruption, migration, low income, budgetary limitations and other factors, this chapter presents further reasonable approaches to finding the solutions to assist Jamaica to overcome its productivity challenges and to serve as an example to other countries who are caught in a similar circumstance.

10.1 Specialisation to Achieve Efficiency

The first thing to realise is that productivity growth has to be engineered. Henry Ford, the pioneer of the industrial revolution, introduced the concept of the production line to increase the productive output of his car manufacturing company. Prior to Ford's invention, it took mechanics 12 hours to fully assemble a car.[3] The process of moving one labourer from one task to the next was time-consuming and reduced the amount

[1] 2018–2021 Medium Term Socioeconomic Plan Planning Institute of Jamaica (2018) https://www.pioj.gov.jm/wp-content/uploads/2019/08/MTF-2018-2021-March-2019.pdf.

[2] Timmer, M. P., Inklaar, R., O'Mahony, M., & Van Ark, B. (2011). Productivity and economic growth in Europe: A comparative industry perspective. *International Productivity Monitor*, (21), 3.

[3] History (Nov, 2021). Ford's Assembly line starts rolling https://www.history.com/this-day-in-history/fords-assembly-line-starts-rolling#:~:text=On%20December%201%2C%201913%2C%20Henry,one%20hour%20and%2033%20minutes.

of output per unit of time. Ford realised that if workers were to focus on a single task and were able to specialise in this task only, they would learn their task better and could complete these tasks in less time. Less time on a single job meant that more jobs could be completed in the same time frame leading to greater productivity. Ford's innovative assembly line took an hour and a half to fully assemble a car, which extended the firm's intensive margin. Eight (8) cars could now be produced in the time it took to produce just one (1). This formula for efficiency has been copied and pasted in the production of many of the world's most mass-produced goods. Although no one person was producing a single good by themselves, the collective effort led to more output per person which increased productivity.

Based on a similar analogy to Ford's industrial line, Smith (1776) posited that it is of the division of labour that the economy achieves its utmost level of productivity. Using the example of making pins, Smith (1776) showed the benefits of specialisation. Four men working individually could make 1000 pins per day each, adding up to 4000 pins in total, while 4 men working in a productive sequence could make 48,000 pins per day. As simple as a pin is, it requires several processes, which when all carried out by one individual makes it less efficient than if each process was carried out by a different individual. Smith (1776) highlighted three characteristics of the Division of Labour that increases productivity: 1. the increase in the dexterity of every workman. The more you do something the better you are at it, 2. time saving: it is time-consuming when a worker moves from one aspect of a task to another and 3. to invent machines that make the work easier. Smith (1776) "The Wealth of Nations" discusses how countries and firms use the principles of economics and the division of labour to create wealth.

Micro, Small and Medium Enterprises[4] (MSMEs) are classified as drivers for growth in Jamaica, (Ministry of Industry & Investment and Commerce, 2022).[5] These firms by definition are small, and many of

[4] Micro businesses in Jamaica employ less than 5 workers and bring in an annual revenue of less than J$15 million. Small and medium businesses employ 6 to 20 and 21–50, respectively. In terms of income, small businesses generate between 15 to 75 million J$, while medium enterprises generate between 75 and 425 million J$. See https://www.miic.gov.jm/content/msme-sector for further details.

[5] Ministry of Industry & Investment and Commerce Agriculture and Fisheries in Collaboration with National Stakeholders (2017). Micro, Small & Medium Enterprises

these enterprises are sole trader, performing most of the tasks themselves, switching concentration from task to task and losing focus in between, thereby achieving the exact opposite of what Ford and Smith proposed. Assume output is positively related to each task but negatively related to the time taken to switch from one task to the next. To maximise the firm's profit, it is rational to minimise the number of tasks that each person must switch to and from. In the case of Ford's factory line, it would be more efficient to have one team design the car and another team make the car instead of the same team both designing and making the car. Those who design the car might not be inclined to make and those who make might not be inclined to design. As a result, division of labour is the more economical approach. Time used to teach the designers how to put together parts or to teach the mechanics how to design could be spent making those who design better designers and those who make become better mechanics.

The idea of specialisation was also proposed by Author Lewis (1963) after World War II in shifting from the traditional agrarian economy to modern industrialisation. This required the movement of labour from rural agricultural sectors where the supply of labour was plentiful to urban areas. Through the process of industrialisation, they would garner a skill or a profession and specialise. This led to the establishment of the Free Zone Areas. The factory production now drives China's global dominance in production and GDP growth. Jamaica and many other SIDS tried to do the same but many factors prevented similar industrialisation from occurring, see Chapter 10.[6]

Looking at the keyboard, have you ever asked yourself why the letters are not placed in alphabetical order? One would expect the letters to be in alphabetical order. When the first typewriter was introduced by Remington in 1874, it was expensive and broke down often because the letters were placed in alphabetical order and the use of certain letters together, for example "s" and "t", jammed the keyboard. Over time, they began to separate the letters and Remington improved. The second typewriter that was introduced in 1878 was better and more fluent.

(MSME) and Entrepreneurship Policy https://www.ceintelligence.com/files/documents/MSME%20and%20Entrepreneurship%20Policy%20Updated%202017.pdf.

[6] See in Chapter 10.

People could type more words per minute which increased their productivity. Continuous understanding of how to improve the purpose that the product serves to the consumer is important if the consumer is to get maximum utility from their purchase. Most industrial countries put effort into expanding the production of goods and services that they have a relative competitive advantage compared to other countries. Swiss cheese, French wine, Italian-designed clothing and English leather are examples of specialisations that have led to internationally distributed goods. Through specialisation in these goods with competitive advantage, these countries have been able to make better use of research and development which increases the amount of output per person employed or output per number of inputs. In order for a country to increase its productivity, it must find what it is good at relative to the rest of the world and direct efforts to make these production processes more efficient. Through this process of constant improvement in efficiency, the country can increase its level of productivity.

10.2 Greater Support Needed for the National Productivity Organisation

To promote productivity growth and boost living standards in the long run, public policies need to focus on improving incentives, capabilities and flexibility within an economy (Banks, 2015).[7] This can be achieved through the implementation of a movement involving the entire population for a decade or more to transform the popular mindset towards hard work, teamwork and creativity, see Prokopenko (1999)[8] and JICA & GRIPS (2011).[9] Particularly, the movement for quality and productivity improvement will require a national effort of many public and private stakeholders to attain economic and social progress. Many countries around the world have invested in National Productivity Organizations (NPOs) to shift the mindset of the people, increase

[7] Gary Banks (November, 2015) Institutions to Promote Pro-Productivity Policies: Logic and Lessons https://www.oecd.org/economy/Institutions-to-promote-pro-productivity-policies-OECD-productivity-working-papers.pdf.

[8] Prokopenko, J. (Ed.). (1999). *Productivity promotion organizations: evolution and experience*. ILO.

[9] JICA represents Japan International Cooperation Agency while GRIPS is the Graduate Institute for Policy Studies in Japan.

productivity, economic prosperity and the standard of living of their citizens (Ramkissoon, 2020).[10]

Between 1950 and 1980, Japan, Korea and Singapore, for example, launched productivity improvement initiatives which enhanced their productivity and competitiveness. Korea launched the Saemaul Movement that significantly improved the productivity among Korean villages. Singapore's productivity movement encapsulated the entire economy including the transportation sector, even taxi drivers had regular conversations about productivity (JICA & GRIPS, 2011). Japan launched a Kaizen movement which included creating Quality Control Circles (QCCs). In Japan, workers were lazy, short-sighted and hardly productive in the early twentieth century (Ministry of Agriculture & Commerce, 1901).[11] Additionally, Japanese had no regard for company rules and executive orders but through the efforts of private firms and public policies, the country was able to transform its "lawless" workers into kaizen workers in half a century (JICA & GRIPS, 2011). Kaizen was a methodology used to bring about structure and facilitate continuous improvement in worker productivity and output. It is an iterative process with 6 steps: (1) identifying problems, (2) analyse current processes, (3) create solutions, (4) test the solutions, (5) measure and analyse results and (6) standardised the solution. This helped to transform Japan's economy into what it is today. In the 1990s, other countries like Burkina Faso and Botswana established their own movements with the assistance from either Japan or Singapore along with the help of multilateral financial agencies. Contemporarily, countries such as India, Mauritius and South Africa have benefited from the establishment of NPOs. As it relates to the Caribbean, the International Labour Organization (ILO) assisted in the establishment of NPOs in St. Lucia and Jamaica in the late 1980s and early 2000s (Ramkissoon, 2020). However, the success achieved by Japan, Korea and Singapore has been elusive for Jamaica. Even the labour productivity of St Lucia is estimated to be an average 42% higher than that of Jamaica.[12]

[10] Mariana Ramkisoon (2020), National Productivity Organizations: repositioning for relevance and impact https://www.ilo.org/wcmsp5/groups/public/---americas/---ro-lima/---sro-port_of_spain/documents/publication/wcms_762768.pdf.

[11] Ministry of Agriculture and Commerce (1901). The survey of industrial workers (in Japanese).

[12] Using data from the World Development Indicators from 1991–2020.

Jamaica's NPO is the Jamaica Productivity Centre (JPC) which was formed in 2003 (approximately 2 decades ago) as a department within the Ministry of Labour and Social Security (MLSS). It is a tripartite of the Government of Jamaica (GOJ), the Confederation of Trade Unions (JCTU) and the Jamaica Employers' Federation. JPC. Ramkissoon (2020) states that the JPC gets approximately 0.60 million US dollars (2019–2020) annually to carry out its functions.[13] This is less than half the funding received by similar NPOs in other countries.[14] The JPC's typical source of funding comes from government subvention from the MLSS's budget, donations, sponsorship from international partners and earning received from services provided. In terms of human capital, the organisation is operating with approximately 47% (7/15) of its max staff complement. The centre is responsible for promoting and facilitating productivity improvements in Jamaica (JPC, 2022).[15] However, many Jamaicans are unaware of the entity due to their low visibility and low national impact, which has to change. The productivity centre should be at the forefront of all discussions related to the transformation of Jamaica from third world to first. Given productivity's importance for sustainable long-term growth, it begs the question of why the national productivity organisation isn't playing a more active role in the coordination of industry activity and business development.

For Jamaica to develop, the island must be engaged in a national campaign to transform people's values, mindsets and aspirations. Many times, serious issues are overlooked, and citizens do not demand the best for themselves. For example, there are a plethora of cases of mismanagement or misappropriation of public funds which causes uproars when highlighted by the media. However, this is short lived as citizens quickly move on without the culprits being held accountable and face the associated consequences. Too often, piece-meal programmes and plans are set in motion without the planned results being achieved. A shift in mindset is necessary or the same mistakes will continue generations after generations to no avail and each subsequent generation will become poorer and

[13] This is based on 2019-2020 funding.

[14] These are the latest annual funding based on work of Ramkisoon (2020): Botswana (US $3.1 million, 2014), Mauritius(US $1.4 million, 2018), South Africa (US $5.96 million, 2019 and India (US $7.50 million, 2018).

[15] Jamaica Productivity Centre (January, 2022) About Us https://www.mlss.gov.jm/departments/jamaica-productivity-centre/.

poorer. JICA & GRIPS (2011) notes that if mindset change does not come spontaneously from the private sector, the state may have to force it from the top until it becomes part of the national culture. Recall how the Kaizen methodology transformed Japan in less than a century.

JICA & GRIPS (2011) noted that based on the experience of Japan, Korea, Singapore and other national productivity movements, the following 6 factors are crucial in the successful transformation of the mindset of a people:

1. Strong personal commitment of the top leader
2. Establishment of core organisations responsible for quality and productivity improvement (such as national productivity organisations)
3. Supporting institutions and mechanisms at central and local levels
4. Massive campaign for mass participation
5. Authorised and standardised training programmes and materials for those concerned
6. Developing private sector capability, especially fostering expertise of private productivity management consultants.

In this context, the leaders of Jamaica must be committed to harnessing the potential of the population.

10.3 Mandatory Second Language

Language in the twenty-first century is likely to become as important as technology was in the twentieth century (Gandara & Acevdo, 2016).[16] This has become even more critical due to aspects of the Fourth Industrial Revolution being fast-tracked by the COVID-19 pandemic. As a result, a greater number of people are now working remotely and participating in the gig economy which grew by 33% in 2020, where the US alone accounted for 2 million new gig workers.[17] Jamaica and many other SIDS who are monolingual can realise significant and well-needed economic

[16] Gandára, P., & Acevedo, S. (2016). Realizing the economic advantages of a multilingual workforce. *UCLA Civil Rights Project. Retrieved on December*, *18*, 2016.

[17] Will The Gig Economy Become The New Working-Class Norm? (forbes.com).

opportunities if they are able to harness the power of becoming multilingual states. Grin et al. (2011) quantitative assessment of the influence of multilingualism on economic processes revealed that foreign language skills can make a considerable contribution to a country's GDP.[18] Correspondingly, Foreman-Peck and Wang (2014) showed that the UK which is predominantly a monolingual country is estimated to lose out on the equivalent of 3.5% of GDP every year.[19] This would amount to a loss of 35% of GDP over a decade if compounded annually, ceteris paribus. A study of small- and medium-size enterprises in Europe including Sweden, Denmark and Germany indicated that companies which invested more in languages were able to export more goods (WEF, 2018).[20] Additionally, German companies that invested heavily in multilingual talent added 10 export countries to their markets while countries who failed to invest or invested less missed out on invaluable contracts. Recent studies have also suggested various trading blocs such as the European Union (EU), the Common Market for Eastern and Southern Africa (COMESA), the Economic Community of West African States (ECOWAS) and some members of BRICS (South Africa and China) as having adopted multilingualism as a strategy of addressing the language factor in their bilateral trading matters.[21] Micheal Schutzler of Forbes Insights also wrote that: "Brazil, Russia, India, China and a host of European, Latin American, and Asian nations—are producing expert executives with outstanding resumes and multilingual fluency."[22] Then the question must be asked. Is this beyond Jamaican and other monolingual SIDS?

[18] Grin, F., Sfreddo, C., & Vaillancourt, F. (2011). *The economics of the multilingual workplace*. Routledge.

[19] Foreman-Peck, J., & Wang, Y. (2014). The costs to the UK of language deficiencies as a barrier to UK engagement in exporting: A report to UK trade & investment. *Cardiff Business School*. The Costs to the UK of Language Deficiencies as a Barrier to UK Engagement in Exporting: A Report to UK Trade & Investment (publishing.service.gov.uk).

[20] World Economic Forum (Feb, 2018) Speaking more than one language can boost economic growth Why speaking more than one language can boost economic growth | World Economic Forum (weforum.org).

[21] Muindi, T., & Isheloke, B. E. (2020). The Role of Multilingualism as a tool in Spurting Trade among the BRICS: An Economic Perspective. *Brics and Economic Development: A Multidisciplinary Perspective*, 113.

[22] Schultzler. (May, 2011). Language Learning Helps Businesses Compete Globally Language Learning Helps Businesses Compete Globally (forbes.com).

Jamaica is strategically located in the Latin America and Caribbean region but the country has not managed to take full advantage of the larger untapped markets in Latin America which could prove useful in bringing new investments, labour market opportunities and an overall boost to economic activity in Jamaica. Why is this so? There is a language barrier. Countries in the Latin American region are predominantly Spanish speaking excluding Brazil whose official language is Portuguese. This has been largely influenced by the history of colonisation by Spain and Portugal in the fifteen and sixteenth centuries which brought forth a unity of all the countries in Latin America.[23] Jamaica was also a country under Spanish colonisation but was later captured by the British in the Western Design in 1655. The Western Design was a plan to end the Spanish dominance in the Spanish West Indies which was orchestrated by Oliver Cromwell and executed by admiral William Penn and lieutenant-colonel Robert Venables. The Western Design was a failure to some extent, but the British ended up with Jamaica as a consolation. The capture of Jamaica by the English was the starting point for the language barrier that Jamaica needs to break to access the larger markets in Latin America today and possibly other Spanish markets worldwide. Jamaica is surrounded predominantly by Spanish countries. To the north of Jamaica you will find Cuba, to the north east Hispaniola (Haiti and Dominican Republic), to the south Panama and Colombia, and to the west Central American countries like Guatemala and Mexico and even Belize which is an English-speaking country but is largely influenced by the Spanish. In essence, if you arrive in Belize as a Spanish speaker with low or no competencies in English, you can still survive. While the influence of the Spanish is still evident across Jamaica in terms of the naming of rivers and streets, the architectural design in some major towns and the old capital of the island (Spanish Town) among others, adopting the language may have greater benefits and may help to bolster Jamaica's economic position over time especially as it relates to bilateral trading arrangements and the flow of capital and labour.

A mandatory inclusion of Spanish as a second language for Jamaica could open up new doors and possibilities that the country can explore to grow the economy, increase productivity and by extension better the lives of Jamaicans. Based on the World Bank's classification, there are

[23] Escobar, Anna María. (2012). Bilingualism in Latin America. https://doi.org/10.1002/9781118332382.ch29. ResearchGate.

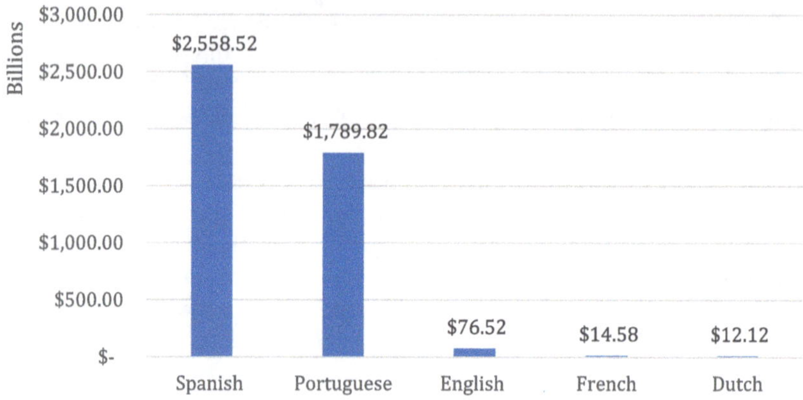

Fig. 10.2 Average GDP in LAC by official language (2010–2020)

42 economies in the Latin America and Caribbean region.[24] Most of the economies are Spanish speaking (43%) followed by English(40%), Dutch(10%), French (5%) and Portuguese (2%).[25] While there is not a large disparity between the number of Spanish and English-speaking territories, the Spanish-speaking countries account for the largest portion of the LAC economy. Over the 2010 to 2020 decade, the average GDP for Spanish speaking territories combined was estimated at US $ 2.55 trillion compared to $76.52 billion for English-speaking territories (see Fig. 10.2). This level of income outstripped all the other countries combined that spoke a different language in region. Income level in Spanish speaking territories was greater than that of the Portuguese, English, French and Dutch by 43, 3,244,17,451 and 21,015%, respectively. Overall, the Spanish represents 57.5% of the GDP in the LAC while the English accounts for a meagre 1.7% (see Fig. 10.3) which presents a case or opportunity for Jamaica that can be exploited.

[24] World Bank Group (2022). World Bank Country and Lending Groups World Bank Country and Lending Groups—World Bank Data Help Desk.

[25] No data was available for British Virgin Islands, St. Martin(French part) and Venezuela. However, they were still included in the total number of countries that were in the LAC region but were not taken into account when the averages were being constructed.

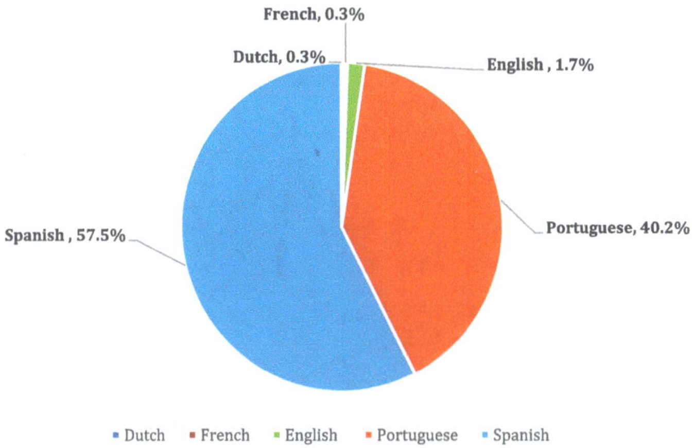

Fig. 10.3 Average share of GDP in LAC by official language (2010–2020)

Presently, most Jamaicans can only compete for local jobs and those that are within Caricom. Their alternative is to migrate to other English-speaking countries in North America or the UK as discussed in Chapter 6. Nonetheless, there are a few Jamaicans who may be at an advantage especially those that are multilingual and work in the BPO industry where they can operate on several different accounts due to their competencies in both English and Spanish. This is more of an exception than the rule. Jamaica must try harder to access the Spanish markets by learning the language of the Spaniards as other countries are making efforts to break barriers that limit international trade with countries that are even remote to them, for example China and Japan have been attracting English teachers from across the globe. Many Jamaicans have also been recruited and have made the journey halfway around the world to assist these Asian countries in becoming better English speakers where the effect is likely to be felt in the generations to come.

Similar to the Asian and European countries that have invested in assimilating other languages and have added new products and export markets to increase trade and economic advancement, so should Jamaica. The island has predominantly been a net importer of goods and service for close to 6 decades which is effectively most of its post-independence era as shown in Fig. 10.4. More critically, Jamaica's exports seem to have been adversely impacted by the 2007–2009 Global Financial Crises and

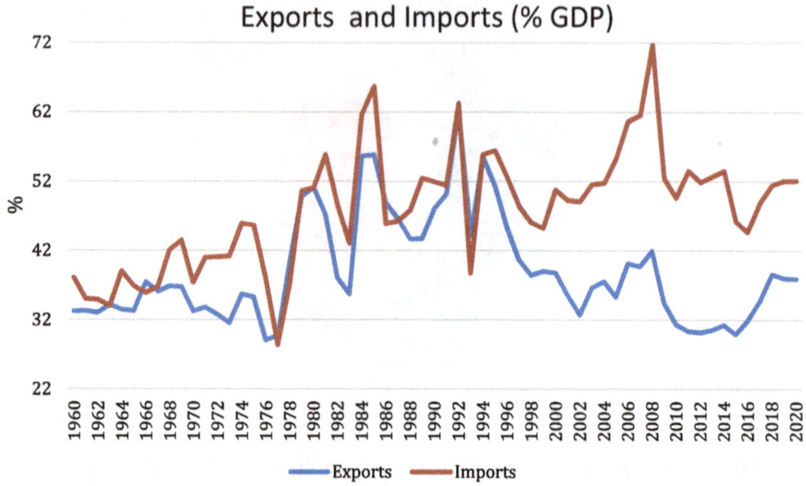

Fig. 10.4 Jamaica's exports and imports (%GDP)

was on a path to recovery when the COVID-19 pandemic ravished the global economy. Figure 10.5 shows that Jamaica's exports fell from US $5.3 billion in 2008 to US $ 4.6 billion in 2020 and imports fell from 9.5 billion in 2008 to 6.8 billion in 2020. Overall, Jamaica's total trade has deteriorated. Countries that are open to international trade tend to grow faster, innovate, improve productivity and provide higher income and more opportunities to their people (World Bank, 2018).[26] In Jamaica's case, the introduction of a mandatory second language could be a strategy to further lower barriers and facilitate more trade. Kavoussi (1984) posits that in low- and middle-income countries like Jamaica export expansion is associated with better economic performance and an important factor of this association is the favourable impact of exports on total factor productivity.[27]

[26] World Bank (April, 2018). Stronger Open Trade Policies Enable Economic Growth for All. https://www.worldbank.org/en/results/2018/04/03/stronger-open-trade-policies-enables-economic-growth-for-all.

[27] Kavoussi, R. M. (1984). Export expansion and economic growth: Further empirical evidence. *Journal of development economics*, *14*(1), 241–250. https://www.sciencedirect.com/science/article/abs/pii/030438788490052X.

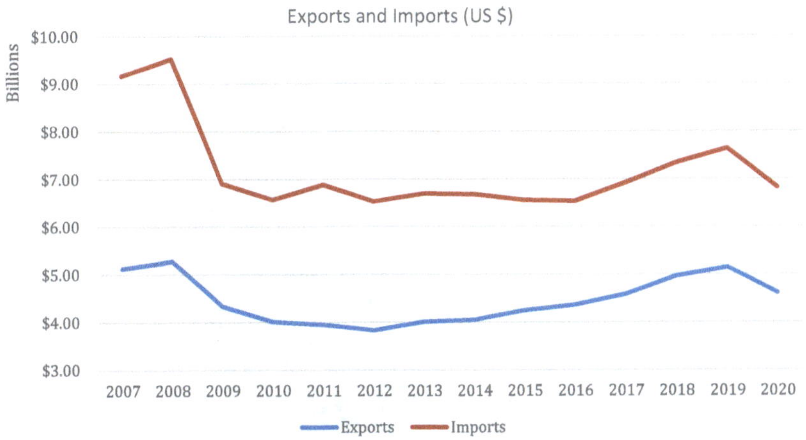

Fig. 10.5 Jamaica's exports and imports (US)

A decomposition of Jamaica's exports to countries provide greater insights into the opportunities that Jamaica could benefit from by introducing Spanish as a second language. Based on data from the Observatory of Economic Complexity (OEC), in 2019 more than 80% of Jamaica's exports were to North America, Europe and Asia.[28] As shown in Fig. 10.6, the United States (32%), Netherlands (10.8%) and Germany (9.4%) accounted for over half of these exports. Interestingly, the Spanish-speaking LAC countries received only 0.6% of the total exports from Jamaica. This could be improved if Jamaica overcame the language barrier. Some Jamaican businesses have already started to take advantage of the Spanish market in the LAC especially those that are in the banking sector. Jamaica Money Market Brokers (JMMB) have several outlets in the Dominican Republic and Sygnus Credit Investments (SCI) have acquired Acrecent Financial Corporation in Puerto Rico and are eyeing investments prospects not only in English Caribbean but also in the Dutch-

[28] These countries included the United States, Netherlands, Germany, Canada, Iceland, Russia, Norway, Azerbaijan, UK and China.

Fig. 10.6 Jamaica's export destinations (2019) (*Source* Organisation for economic cooperation and development)

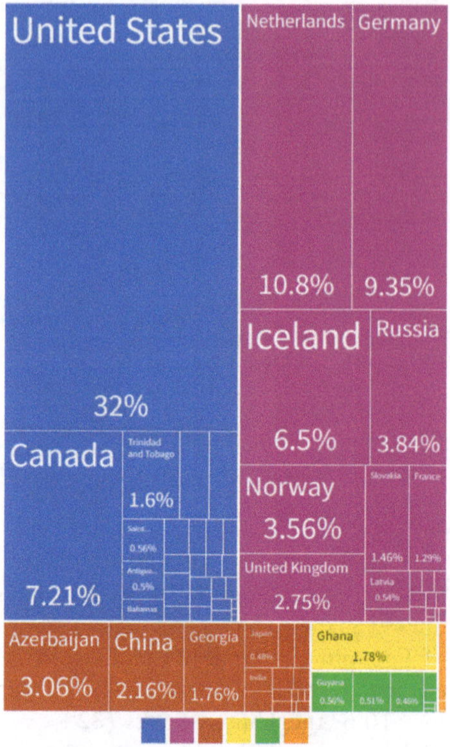

and Spanish-speaking Caribbean.[29] Other companies in different industries can follow to increase their own earnings and also boast the exports of the country.

10.4 Increase Economic Complexity

Economic complexity reflects the knowledge embedded in the productive structure of a country. Hidalgo and Hausmann (2009) posited that economic complexity is strongly correlated with income per capita and predictive of future economic growth and exports. Stojkoski and Kocarev

[29] Gleaner (Februrary,2022) Sygnus revise growth strategy, eyeing share buy-back plan https://jamaica-gleaner.com/article/business/20220218/sygnus-revises-growth-strategy-eyeing-share-buy-back-plan.

(2017) research also gave support to Hidalgo and Hausman (2009) as their research shows that economic complexity is a statistically significant explanatory variable of growth in the long run, and thus, it creates enormous economic implications for the development of a nation.[30] Balland et, al. (2022) notes that alterations in an economy's productive structure can be understood as an amalgamation of two factors[31]: (1) the process by which countries find new products and unexplored combinations of the capabilities they already have and (2) the process by which countries accumulate new capabilities and combine them with other previously available capabilities to develop more products.

Over the last 2 decades, Jamaica's economy has become relatively less complex, moving from a rank of 79th to 83rd on the Economic Complexity Index (ECI).[32] This also coincides with the declining trend in labour productivity and the low human capital base of the country, see Chapter 5. Relative to other countries in the Caribbean region, Jamaica seems to have one of the more unsophisticated economies as Fig. 10.7 shows that Jamaica is only more complex than Guatemala, Honduras and Cuba. In order for Jamaica to become more complex and increase its rank in the region, investments in human capital and technology will have to be prioritised. This will allow Jamaicans to create higher value-added products such as microchips, softwares and more advance technological improvements in industries such as agriculture and manufacturing.

By understanding that productivity growth is a process which must be engineered, the productivity centre must play a crucial yet challenging role in helping Jamaica to achieve its productivity objectives. A useful strategy is the development of a road map that will ultimately guide the country towards institutionalizing processes that will form the backbone of a national productivity policy used for standardised processes. Spanish

30 Stojkoski, V., & Kocarev, L. (2017). The relationship between growth and economic complexity: evidence from Southeastern and Central Europe. https://ideas.repec.org/p/pra/mprapa/77837.html#:~:text=The%20index%20of%20economic%20complexity,and%20the%20products%20they%20export.&text=The%20results%20show%20that%20the,it%20creates%20enormous%20economic%20implications.

31 Balland, P. A., Broekel, T., Diodato, D., Giuliani, E., Hausmann, R., O'Clery, N., & Rigby, D. (2022). The new paradigm of economic complexity. *Research Policy*, *51*(3), 104450.

32 OEC. (February, 2022). Economic Complexity Ranking https://oec.world/en/profile/country/jam?compareEciSelector=compareBy2&compareExports0=comparisonOption5.

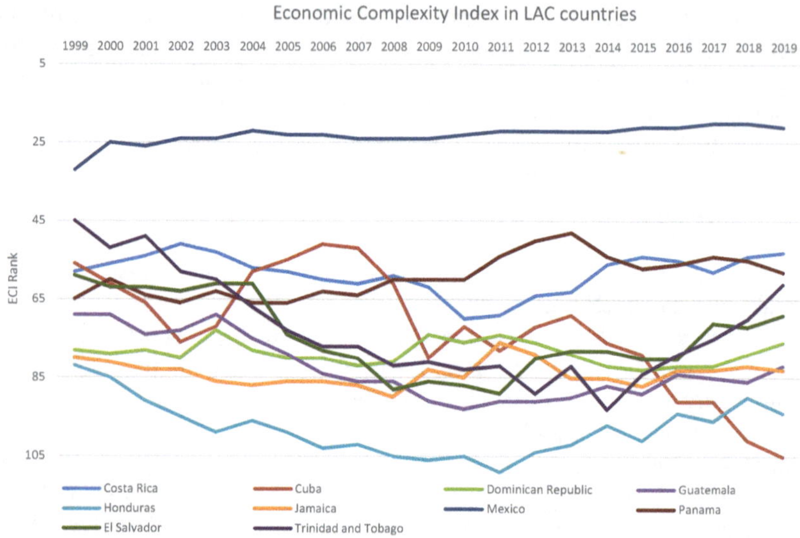

Fig. 10.7 Economic complexity of LAC countries (*Source* Organisation for economic cooperation and development)

is already taught in many schools in Jamaica. This makes it relatively easy to set annual targets to improve the ratio of the Jamaican population that has competencies in reading, writing and speaking basic Spanish. If these suggestions are implemented, it is likely that the improved characteristic of the Jamaican economy will help to drive economic complexity in the right direction as the island moves to a more prosperous nation where people will want to live and even those that have already migrated would consider coming back to Jamaica.

10.5 How to Escape the Productivity Trap?

Latin American and Caribbean countries face severe challenges in their attempt to meet the evolving needs of individuals and businesses. In this ever-changing world, business demands are rapidly changing with technology and staffing requirements must progress to fulfil these functions so that businesses can progress. If not, then a country will aspire but never achieve the level of productivity growth and standard of living it desires. Many countries have found themselves in four fundamental traps that have presented challenges to materializing the desired levels of economic growth and developmental according to the Latin American Economic Outlook 2019. These include a productivity trap, social vulnerability trap, institutional trap and environmental trap. Many are characterised by widespread poverty and a very vulnerable middle class because of innate challenges that are embedded within.

How severe is the social vulnerability trap?
In many Latin American and Caribbean countries, the middle class remains vulnerable and face the risk of becoming poor once more due to innate structures of the economy. Many college graduates in these countries have to settle for low-quality jobs that do not afford them the ability to save and invest to increase their future earnings to create a buffer against poverty and adverse economic shocks. For example, many college graduates can only find jobs in their degree area or any degree area and ends up working for subsistence income in a call centre in the BPO industry. Lack of investment in key social services, for example, health care and garbage disposal etc. reduces the level of social protection and the standard of living. Unless these innate challenges are addressed where college graduates can earn an income that allows them to save and invest, these countries will be trapped in a low output low productivity cycle.

What is the institutional trap?
Institutions in these countries are failing to respond to the needs of its citizens. According to the Latin American Economic Outlook 2019, this creates low satisfaction and distrust in critical processes that these countries need to be sustainable. The widespread corruption and favouritism in personal and business processes has weakened citizens' confidence. Citizens place less value on core national duties, for example paying taxes. Furthermore, in many of these countries key civil servants are not compensated enough to motivate them to carry out their functions effectively. Nurses, doctors, teachers and police are underpaid and therefore demotivated. Many of the best migrate which contributes to low absorption rate that reduces the level of productivity more.

What is the environmental trap?
Latin America and Caribbean countries depend mainly on export that depletes natural resources. These are becoming greater and poses a risk to the sustainability and long-run viability of these business ventures. The

global climate change policy will require the need to redirect many of these processes to more stable long-term alternatives that will require a path towards developing a more sophisticated economy.

Overcoming the productivity trap.
Latin American and Caribbean countries are experiencing low productivity and slow growth in many sectors because their productive sectors have not evolved to incorporate technological advancements in the procedures and processes that increase worth. According to the Latin American Economic Outlook, these countries largely produce and export primary production or extractions which are not sufficiently vertically or horizontally diversified to create value-added products. The value of what you produce gives an indication of the level of productivity. If the value of what you produce does not increase, it means productivity has not increased. Service sectors are also sustained in primary lower-level category. These domestic linkages must improve to increase value creation through value-added diversification.

Index

A. Haughton and W. Ivey, *Overcoming Productivity Challenges in Small Countries*, https://doi.org/10.1007/978-3-031-23301-2

C

D

R

S

Kolibri 360 GmbH
Ferdinand-Jühlke-Straße 7
99095 Erfurt, Deutschland
produktsicherheit@kolibri360.de